RAND NATIONAL DEFENSE RESEARCH INSTITUTE

T0122920

Satellite Anomalies

Benefits of a Centralized Anomaly
Database and Methods for Securely
Sharing Information Among Satellite
Operators

David A. Galvan, Brett Hemenway, William Welser IV,
Dave Baiocchi

Prepared for the Defense Advanced Research Projects Agency

Approved for public release; distribution unlimited

This research was sponsored by DARPA and conducted within the Acquisition and Technology Policy Center of the RAND National Defense Research Institute, a federally funded research and development center sponsored by the Office of the Secretary of Defense, the Joint Staff, the Unified Combatant Commands, the Navy, the Marine Corps, the defense agencies, and the defense Intelligence Community.

Library of Congress Cataloging-in-Publication Data

ISBN: 978-0-8330-8586-3

The RAND Corporation is a nonprofit institution that helps improve policy and decisionmaking through research and analysis. RAND's publications do not necessarily reflect the opinions of its research clients and sponsors.

Support RAND—make a tax-deductible charitable contribution at www.rand.org/giving/contribute.html

RAND® is a registered trademark

Cover image courtesy of NOAA

© Copyright 2014 RAND Corporation

RAND OFFICES
SANTA MONICA, CA • WASHINGTON, DC
PITTSBURGH, PA • NEW ORLEANS, LA • JACKSON, MS • BOSTON, MA
CAMBRIDGE, UK • BRUSSELS, BE
www.rand.org

Preface

The Defense Advanced Research Projects Agency (DARPA) developed the Programming Computation on Encrypted Data (PROCEED) program to improve efficiency of algorithms that carry out computations on encrypted data without having to decrypt the data themselves. These algorithms would allow multiple users to contribute proprietary data to perform analysis without worry that the data themselves will be exposed, either to each other or to a third party. RAND investigated the concept of a secure, shared database of satellite anomalies as a potential use case for DARPA's PROCEED program. In particular, we were interested in whether such cryptographic techniques as secure multiparty computing could contribute to enabling a securely shared anomaly database that would be useful to the satellite operator community.

Satellite anomalies are mission-degrading events that affect on-orbit operational spacecraft. Every satellite experiences anomalies at some point during its operational life cycle, with degrees of severity ranging from gradual degradation of solar panel efficiency to complete and sudden loss of contact or mission failure. Most satellite owners keep track of satellite anomalies for their own systems, but many closely guard anomaly data from disclosure to others. Those tasked with investigating the causes of severe or recurrent anomalies would benefit from greater awareness of whether other satellites are also experiencing anomalies while under similar conditions.

This report describes the nature and causes of satellite anomalies and discusses the potential benefits of a shared database of anomalies for use by the satellite operator community. It also suggests cryptographic methods that may encourage the secure sharing of satellite anomaly information while addressing the concerns of those organizations that consider their satellite information proprietary. The findings presented here should be of interest to government agencies working with satellite data (especially the National Aeronautics and Space Administration and the National Oceanic and Atmospheric Administration), the Department of Defense, federally funded research and development centers, and satellite owners and operators in the commercial sector.

This research was sponsored by DARPA and conducted within the Acquisition and Technology Policy Center of the RAND National Defense Research Institute, a federally funded research and development center sponsored by the Office of the Secretary of Defense, the Joint Staff, the Unified Combatant Commands, the Navy, the Marine Corps, the defense agencies, and the defense Intelligence Community.

For more information on the RAND Acquisition and Technology Policy Center, see http://www.rand.org/nsrd/ndri/centers/atp.html or contact the director (contact information is provided on the web page).

Contents

Figures and Table

Figures

Table

Summary

Satellites in Earth orbit represent a critical component of society's modern infrastructure. Failure of a satellite, or even of a particular subsystem, can significantly affect capabilities on which the civil, defense, and commercial sectors have come to rely. Most satellite owners monitor the health of their systems and keep track of any unusual problems their spacecraft experience. These "satellite anomalies" are defined, for the purposes of this study, as any mission-degrading events affecting on-orbit operational spacecraft. Examples include onboard computer errors or failures, malfunctioning attitude control systems, loss of radio contact, the degradation of solar panel efficiency, and many other mechanical or electronic symptoms. Most satellites experience anomalies of varying severity throughout their lifetimes. The root causes of these anomalies may include manufacturing and design flaws in satellite hardware and software, extreme space weather events that affect the intensity of electromagnetic radiation and density of charged particles in the satellite's environment, impacts with micrometeoroids or space debris, operator error, regular wear and tear from exposure to the plasma environment of space, or interference by human technological activities, either intentional or unintentional.

Since many threats to a satellite's mission first manifest themselves as unexplained anomalies in the telemetry, operators tend to track these anomalies in their own databases of detailed satellite status information, which many operators consider proprietary or classified. They may make efforts to investigate the cause of these anomalies if they are repetitive or significant enough to threaten the satellite's mission. In such investigations, a centralized and up-to-date shared database of anomalies experienced by many different satellites in a variety of orbital configurations could help provide context and narrow down the potential causes of the anomalies. Investigators could use the database to help determine, for example, whether the problem they are experiencing is unique to their satellite or common to many satellites at a given time or in a given region of space. This could help inform whether the cause is a hardware defect, accidental interference, purposeful attack, or a space weather event. However, there are obstacles that inhibit satellite owners from the civil, defense, and commercial sectors from developing such a database or sharing information about their satellite systems with other groups.

This report explores how a centralized satellite anomaly database could benefit the community of spacecraft operators, and how the obstacles inhibiting the development of a shared database could be partially overcome. To address these issues, we have conducted a literature review as well as discussions with subject matter experts on space physics, engineering, satellite anomaly investigation, and the insurance industry.

Our analysis resulted in the following observations:

- **A centralized and standardized satellite anomaly database is recognized by subject matter experts from the organizations we contacted—including the National Oceanic and Atmospheric Administration (NOAA), the Aerospace Corporation, and commercial companies—as a potentially valuable resource for the satellite operator community.** Such a database would aid in anomaly investigations, hence reducing costs and increasing efficiency. It could also contribute to the scientific understanding of the real-world impacts of the near-Earth space environment.[1]
- **A single centralized database could offer advantages over multiple smaller ones.** Multiple smaller databases currently exist, but

---

[1] This would be consistent with the scientific goals articulated by the National Science Foundation's *Geospace Environment Modeling* (2012) and the National Aeronautics and Space Administration Living With a Star program (2012b).

they tend to be either broadly available but incomplete, or highly detailed but not broadly available. Individual satellite operators may maintain their own internal anomaly databases, but these multiple databases vary in accuracy and content. A centralized database could reduce duplication of effort, while providing data that is both detailed and broadly available.

- **The development of a centralized satellite anomaly database that would be useful to the broad satellite community is hindered by concerns about sharing proprietary information and by a lack of dedicated resources for development and maintenance.** Concerns over sharing of proprietary information are perhaps the most significant obstacle for companies in the commercial sector. Commercial satellite operators have also articulated that they might expect an organization like NOAA to provide anomaly database analysis and curation as a government service. However, the lack of resources for development and maintenance is a problem in the civil and defense sectors of government, which would likely include organizations that could serve as trusted third parties or those that could develop the encryption technologies that could help address the privacy concerns associated with sharing anomaly data.

- **Management of a centralized database by a trusted third party, encryption techniques such as secure multiparty computing, and differential privacy may help overcome inhibitions of commercial satellite operators to share anomaly information, thus contributing to greater benefit throughout the satellite operator community.** The application of these techniques to the development of a securely shared satellite anomaly database is a potential use case for the Defense Advanced Research Projects Agency's Programming Computation on Encrypted Data program. One option would authorize a trusted third party to ingest and manage the data provided by contributing satellite operators; to categorize the anomalies; and to reveal information only in aggregate, statistical form to avoid divulging proprietary identifying information from the contributors. Candidate organizations that could serve as a trusted third party include NOAA and certain federally funded research and development centers that have expertise in satellite anomalies and a lack of financial interest in learning or disclosing proprietary information from contributing organizations. Satellite insurance companies could also potentially play this role, as they have an interest in maintaining the data and in keeping them secret for their clients.

Another option utilizes cryptographic methodologies such as secure multiparty computation and "differential privacy," enabling operators to contribute satellite anomaly information to the database without worry of their proprietary information being compromised. This may reduce or even eliminate the need for a trusted third party in managing the database.

A slightly different option allows operators to contribute only what information is necessary to describe elements of the anomaly that would make them minimally useful to the community, while concealing the identity of the satellite, perhaps including only time, type of orbit, and subsystem affected in the anomaly listing.

Though they do not represent a complete solution to the problem of securely sharing anomaly information for the benefit of all sectors of the satellite community, these proposed methods show promise and could be further developed to contribute to a comprehensive solution.

- **Automated "satellite as a sensor" methods for identifying and cataloging anomalies may also reduce the workload of those investigating satellite anomalies.** Such methods rely on pattern recognition algorithms that can automatically (or semi-automatically) recognize and categorize anomalies in a satellite's telemetry data stream.

These observations and suggested methodologies may be useful as organizations consider the benefits available from a shared database, including improved success in anomaly investigation, enhanced space situational awareness, increased efficiency of the aerospace industry, and a better understanding of the near-Earth space environment through empirical observations.

Acknowledgments

The authors would like to thank our subject matter experts for taking the time to hold discussions with us, including: Dr. Janet Green of NOAA's National Geophysical Data Center, Dr. T. Paul O'Brien and Dr. Chris Tschan of the Aerospace Corporation, Ms. Nikki Noushkam of Orbital Sciences Corporation, Dr. Joseph Allen (retired, formerly of NOAA), and David Hoffer of Atrium Insurance Corporation. Thanks also to Holly Johnson, who was very helpful in the organization and editing of this document. Thanks to our peer reviewers, Lindsay Millard and Jan Osburg, for providing critical commentary that helped to improve the report. Thanks also to Paul DeLuca and Cynthia Cook for useful suggestions.

Abbreviations

ASIC	Atrium Space Insurance Corporation
CIR	co-rotating interaction region
CME	coronal mass ejection
DARPA	Defense Advanced Research Projects Agency
DoD	U.S. Department of Defense
EDAC	error-detection-and-correction algorithms
ESD	electrostatic discharge
FFRDC	federally funded research and development center
GEO	geosynchronous orbit
IMF	interplanetary magnetic field
JSPOC	Joint Space Operations Center
K	degrees Kelvin
keV	kiloelectron volt
LEO	low Earth orbit
MEO	medium Earth orbit
MeV	Megaelectron volt
MPC	multiparty computation

NASA	National Aeronautics and Space Administration
NGDC	National Geophysical Data Center
NOAA	National Oceanic and Atmospheric Administration
NRC	National Research Council
PIR	private information retrieval
PROCEED	Programming Computation on Encrypted Data
R_E	Earth radii
SAA	South Atlantic Anomaly
SEAES	Spacecraft Environment Anomalies Expert System
SEE	single event effect
SEU	single event upset
SND	Space News Digest
SSN	Space Surveillance Network
SSPA	solid-state power amplifier
TID	total ionizing dosage
TMR	triple-modular redundancy

CHAPTER ONE

Introduction

Satellite systems have become critical components of infrastructure in the civil, defense, and commercial sectors. In the civil sector, government agencies and research scientists use satellite observations to improve our understanding of the space and terrestrial environments on a grand scale, with scientific targets ranging from the Earth's surface and atmosphere to the space environment of the solar system and the most distant objects in the universe. They also use satellites to inform practical policy decisions in efforts to improve the human condition, with programs to monitor weather, natural and man-made hazards, agricultural development, and the global impact of human activity. The defense sector relies on satellites for communications, remote command and control, global positioning and timing, reconnaissance and intelligence, and environmental monitoring to contribute to national security and give warfighters a strategic advantage. The commercial sector has become heavily dependent on satellites to provide the backbone of global communications. The unique vantage point available from Earth orbit has enabled unprecedented efficiencies in global business collaboration through communication, information distribution, and fast electronic monetary transactions.

A significant portion of the world population has come to both expect and rely upon technologies enabled by the more than 800 operational satellites in orbit today.[1] As our society relies more heavily on

---

[1] The United States Strategic Command uses the Department of Defense (DoD) Space Surveillance Network (SSN) of radar and optical telescopes to detect, track, and maintain a catalog of all man-made objects in Earth orbit through the Joint Space Operations Center

infrastructure in space, we must maintain awareness of the threats to satellite reliability. Part of that awareness involves understanding the phenomena that may result in a partial or complete failure of the satellite and its ability to provide critical services. The vantage point of space provides many benefits, yet the space environment itself can be extremely hazardous to modern electronics.

Many satellites encounter anomalous events detrimental to mission performance at some point during their operational lifetimes. These "satellite anomalies" may be as minimal as a temporary error in a noncritical subsystem, or as devastating as a complete mission failure. Hardware damage and software malfunctions, the typical manifestations of these anomalies, may occur because of a variety of causes, including faulty equipment, the hazardous natural space environment, impact with orbital debris, operator error, hostile actions by a malicious actor, or even unintentional interference from another satellite transmitter. The cause of the anomaly is typically not obvious to the satellite operators at the time of the event.

While individual satellite operators may investigate and catalog their anomalies, few databases are available to the broader satellite community. Those that are broadly available and openly shared are mostly limited to historical anomalies encountered by a small number of scientific satellites whose operators and sponsoring agencies had the resources and willingness to share the information openly. These data are highly valuable for mission design purposes, as they provide an empirical record of which hardware and software designs are most robust, and which regions of space are most hazardous under varying solar-terrestrial conditions.

However, the absence of a centralized, accurate, and up-to-date anomaly database available to the broader community means that satellite operators do not typically have access to information about anomalies other satellites may be experiencing, at similar times, in sim-

---

(JSPOC), which reports that it is currently tracking ~16,000 Earth-orbiting objects, about 5 percent of which are functioning satellites (the rest are inactive satellites, debris, or rocket bodies) (Vandenberg Air Force Base, 2013). Hence, estimates of the number of functioning operational satellites currently in orbit range from ~800, as reported by the JSPOC, to ~1,000, as mentioned by National Aeronautics and Space Administration (NASA, 2011a).

ilar orbits, and under similar conditions. Such information would be useful in diagnosing causes of anomalies. For example, if many satellites experience similar anomalies around the same time and in similar regions of space, it may indicate an environmental hazard as the cause (e.g., a strong natural space weather event such as a solar flare, solar proton event, or magnetospheric storm). If a single satellite experiences an anomaly while no other satellites in the region do, it may suggest a hardware or software problem unique to that satellite or component. And if some satellites experience recurrent problems with a particular hardware component, operators of other satellites using that same component may benefit from awareness of those problems, potentially avoiding some of the investigation and recovery costs if they experience similar anomalies.

A centralized satellite anomaly database may provide mutual benefit among the community of satellite operators. It would enable sharing of current information on anomalies that occur in satellites under a variety of solar-terrestrial conditions and in different regions of near-Earth space. Helping to narrow the possible causes of an anomaly can save time spent investigating, hence reducing costs for the operators and their customers.

Despite these potential benefits, a comprehensive, up-to-date, and broadly accessible database of anomalies from a large number of satellites does not exist. There are political, economic, and operational obstacles to the development and maintenance of such a database. For one, there are distinct disincentives for some satellite operators to disclose anomaly information to a wider community. Satellite operators in DoD often have strict national security requirements prohibiting them from sharing satellite information specific enough to be useful in anomaly investigations. Commercial satellite owners may not wish to reveal to their competitors or investors that their on-orbit assets are experiencing unsolved technical problems. Also, there are logistical efforts required to organize and manage a community service like an anomaly database, with little funding or resources typically available for such a task. Civil space agencies—e.g., NASA, the National Oceanic and Atmospheric Administration's (NOAA)—may be willing to share anomaly information with others but may not have the resources allocated to manage the databases themselves or to develop standards for information sharing.

Despite these obstacles, there is recognition in the satellite operator community that a centralized database could provide considerable value. In April 2012, NOAA's National Geophysical Data Center (NGDC) and Space Weather Prediction Center hosted a workshop in Boulder, Colo., for "Satellite Anomaly Mitigation Stakeholders."[2] The goal of the meeting was to "identify the space weather impacts to satellite infrastructure and define NOAA services needed to mitigate those impacts" (Green et al., 2012a). Among the attendees were representatives from a variety of organizations, including experts on the space environment (NASA and NOAA), and representatives from commercial satellite operators (e.g., Intelsat, Inmarsat), satellite insurance corporations (e.g., Atrium Space Insurance Consortium), and federally funded research and development centers (FFRDCs) working with DoD (e.g., Aerospace Corporation). Along with interest in the availability of NOAA's 1–2 day forecasts of the space environment, the commercial satellite company representatives articulated a strong interest in having access to a community-shared database of anomaly activity, possibly managed by NOAA's NGDC (Green et al., 2012a; Green et al., 2012b).

Several potential models and techniques for managing a centralized satellite anomaly database may help overcome the obstacles inhibiting data-sharing among members of the satellite operator community. In particular, the Defense Advanced Research Projects Agency's (DARPA's) Programming Computation on Encrypted Data (PROCEED) program supports the development of cryptographic techniques that could enable useful operations on encrypted data without decrypting it, with the goal of making such computations practical in an environment where the associated parties may not trust one another. RAND was asked to consider the problem of securely sharing satellite anomaly information among multiple users as a potential use case for the techniques being developed in the PROCEED program.

---

[2]   Materials from the NOAA Satellite Anomaly Mitigation Stakeholders meeting are available by request from NOAA and online (NOAA, undated a).

The study was driven by the following research questions: What are satellite anomalies, and how do they affect the functionality of spacecraft? What phenomena cause them? How can cataloging these anomalies in a centralized database aid satellite designers and operators? What anomaly databases currently exist, and how could such databases be changed to optimize utility for anomaly investigators? What type of information would the ideal centralized satellite anomaly database contain? What are the obstacles preventing a useful anomaly database from being developed? How could such information be shared securely in the interest of protecting proprietary information from exposure?

To address these questions, we conducted a literature review and discussions with subject matter experts on the near-Earth space environment, the nature and causes of satellite anomalies, and the techniques used to investigate and catalog them.

In this report, we begin by discussing the nature of satellite anomalies and their primary causes. While other sources have discussed anomalies caused by the natural space environment in detail, here we offer a broad overview of the wider range of potential causes of anomalies, including the natural environment, faulty hardware and design, man-made interference, and orbital debris. Next, we review existing satellite anomaly databases and how they differ from the proposed centralized database. We provide an overview of the process of anomaly investigation and cataloging conducted by satellite operators or manufacturers. We then discuss methodologies that may facilitate the sharing of satellite anomaly information via a centralized database while satisfying contributors' concerns about information security, proposing three sample solutions involving a trusted third party, secure multiparty computing (MPC), and differential privacy. Finally, we summarize our observations and provide recommendations for future efforts that could enable the existence of a useful centralized anomaly database.

CHAPTER TWO

Satellite Anomalies

The Natural Space Environment

A brief overview of the near-Earth space environment is useful to provide context for the regime in which anomalies occur. The most important player in our space environment is the sun. The sun is a G-type main sequence star with a temperature of nearly 15.7 million degrees Kelvin (K) at its core, ~6,000 K at the photosphere (the sun's surface), and ~1 million K in the corona, the sun's atmosphere. At these high temperatures, all of the sun's material is in the plasma phase, since the outer electrons of individual atoms have enough energy to escape their atomic nuclei, resulting in charged ions and free electrons. Thus, the sun is made of plasma, as is the solar wind, a constant stream of material that moves outward into space, carrying the sun's magnetic field with it. The solar wind consists primarily of protons and electrons with densities ~5 cm^{-3} (particles per cm^3), and typical outward velocities of 300 km/s. These densities and speeds can vary significantly, however, when interactions of complex magnetic fields near the sun's surface produce explosive events such as coronal mass ejections (CMEs) and solar flares. CMEs spew higher-density solar plasma out into space at up to several times the speed of the surrounding solar wind. If they are properly aligned, CMEs can arrive at the near-Earth space environment within several days. Solar flares are massive explosions that emit high-intensity electromagnetic radiation and may also produce accelerated protons. CMEs and solar flares are frequently associated with one another but have also been observed to occur independently. Protons from solar flares are typically less numerous but often of higher energy and quicker to arrive at the Earth than the material in a CME. Whether produced by a solar flare or a CME, the arrival of high-energy protons at near-Earth space is called a solar proton event (Park et al., 2012). There are also "streamers," coronal holes, and "spicules," all of which are regions of open magnetic field lines on the sun where solar wind material flows outward more quickly than in surrounding regions of the photosphere and corona. When this faster solar wind collides with slower solar wind as it moves outward into the solar system, it forms a phenomenon known as a co-rotating interaction region (CIR). A CIR is a region of dynamic, fluctuating particle densities and magnetic field strengths and orientations. Both CIRs and CMEs can interact with the Earth's magnetic field, producing variations in plasma densities and magnetic field strength and orientation in near-Earth space and the upper atmosphere. These variations at Earth caused by solar phenomena are collectively referred to as "geomagnetic activity."

The rate of occurrence of solar events described here is correlated with the abundance of sunspots, which are regions of reduced temperature and enhanced magnetic field strength on the photosphere. Both sunspots and geomagnetic activity levels follow an 11-year solar cycle, with maxima in solar activity separated from minima by approximately 5.5 years (e.g., Wertz and Larson, 1999; Kivelson and Russel, 1995). The most recent solar minimum was in 2008, and solar activity has risen until recently, with an apparent peak late 2013 or early 2014 (NASA, 2014).

Earth has an internally generated magnetic field that extends far beyond its atmosphere, carving out a three-dimensional region of space known as the magnetosphere. Within the magnetosphere, the Earth's approximately dipolar magnetic field is dominant. Outside the magnetosphere, the magnetic field created by the sun and the solar wind, known as the interplanetary magnetic field (IMF), is dominant. Upon encountering the magnetosphere, the charged particles in the solar wind react to the Earth's magnetic field and stream around and through the magnetosphere in complex ways, compressing the magnetosphere on the sunward side, and stretching the magnetosphere out into a "magnetotail" behind the Earth. The dayside magnetopause (the sunward boundary of the magnetosphere) is typically about 10 Earth radii (1 R_E = 6,371 km) from the center of the planet, while the magnetotail continues for hundreds of R_E behind the planet. Solar material enters the magnetosphere through direct streaming along magnetic field lines at the cusps of the magnetosphere near the north and south magnetic poles, allowing a constant stream of solar wind particles at high latitudes to impact the atmosphere and generate the aurora. The solar plasma also enters the magnetosphere both at the dayside magnetopause and at the flanks of the magnetotail at "reconnection sites"—regions where the IMF field lines intermittently connect to the Earth's magnetic field, depending on the character of the solar wind density, speed, and IMF direction (e.g., Wertz and Larson, 1999; Kivelson and Russel, 1995). The quintessential example of geomagnetic activity is the geomagnetic storm, a period of intensified magnetic field strength and injection of high-energy plasma into the Earth's magnetosphere. The main phase of a geomagnetic storm, during which satellite anomalies would be expected to occur, typically lasts between two and eight hours, with a gradual recovery to prestorm conditions that may last multiple days.[1]

Inside the magnetosphere, high-energy charged particles are concentrated in the Van Allen radiation belts, which extend from ~1,500 km altitude out to ~6 R_E. These represent the most consistently present threat of satellite anomalies, especially to satellites in medium Earth orbit (MEO) and, to a lesser extent, geosynchronous orbit (GEO). Satellites in low Earth orbit (LEO), with altitudes up to ~2,000 km, experience high-energy particles from the radiation belts to a lesser degree than satellites at MEO and GEO, though high-inclination LEO satellites are exposed to the dynamic polar regions where geomagnetic activity can cause intense particle streams to concentrate. Figure 2.1 shows a schematic of the Earth's magnetosphere, pointing out the approximate locations of satellites in LEO, MEO, and GEO orbits.

---

[1]   More information on timescales of typical geomagnetic phenomena is available at NOAA (undated b) and at NASA (2012c).

Figure 2.1
The Near-Earth Space Environment

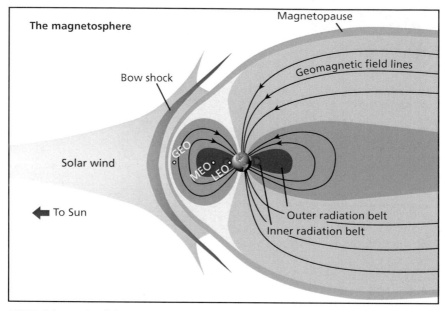

The magnetosphere

Magnetopause

Bow shock

Geomagnetic field lines

Solar wind

GEO

MEO

LEO

To Sun

Outer radiation belt

Inner radiation belt

NOTE: Schematic of the Earth's magnetosphere, with several regions labeled where satellite anomalies often occur, including the Van Allen Radiation Belts (Inner and Outer), and typical satellite orbits (GEO, MEO, and LEO).
RAND *RR460-2.1*

Anomaly Causes and Contributing Factors

For the purposes of this study, we follow the convention of the National Research Council (NRC) and define a *spacecraft anomaly* as a "mission-degrading or mission-terminating event affecting on-orbit operational spacecraft" (NRC, 2011). Most satellites encounter some mission degradation at some point during their operational lifetimes. The spectrum of severity in these degradation events ranges from gradual decrease of a subsystem's efficiency (e.g., solar panel degradation because of ambient low-energy charged particles) to temporary service outage (e.g., electromagnetic interference of communication satellites) to sudden and complete mission failure (e.g., a single event upset [SEU] because of cosmic ray impact).[2] Along that spectrum, many anomalies manifest as abnormalities in telemetry or errors in spacecraft electronics, often because of electrostatic discharge from surface charging, internal charging, or high-energy charged particles imparting their energy into onboard solid-state memory, "flipping a bit" and causing incorrect computations that may affect spacecraft performance (single event effects, or SEE). Such electronic problems may result in mechanical symptoms, such as loss of attitude and pointing control.

One high-profile example of such an anomaly is the failure of the momentum wheel control systems of the Anik E1 and Anik E2 Canadian telecommunications satellites on January 20, 1994, resulting in a temporary loss of data and television service for millions of users throughout Canada. The anomalies were likely caused by an electrostatic discharge in the momentum wheel control systems of both satellites, brought on by internal spacecraft charging from enhanced high-energy electrons because of a coronal hole on the sun (Lam et al., 2012).

Anomalies in general may result from a variety of different causes. The natural space environment provides many phenomena that can affect insufficiently shielded electronics. High-energy particles over 1 megaelectron volt (MeV) may emanate from distant supernovae ("cosmic rays"), the Van Allen radiation belts, or solar proton events, and may contribute to SEEs and internal ("deep dielectric") charging. Mid-range energetic particles in the hundreds of kiloelectron volts (keV) contribute to internal charging and come from the Van Allen belts or as injections into the inner magnetosphere from the plasma sheet during periods of intense geomagnetic activity. Solar flares can produce broadband electromagnetic emissions, including intense X-rays

---

[2] Note that the physical phenomenon causing the degradation is not the "anomaly." "Anomaly" more properly refers to the effects as observed by the satellite operator: a decrease in power, a failure of a subsystem, etc. The cause of these effects will not be immediately known by the operator (though they may suspect a cause), and the event is initially referred to as an "anomaly." The investigation of that anomaly may (or may not) lead to an explanation that partially or fully accounts for the degradation/failure event. In the case of solar panel degradation, the anomaly is the reduction of the solar panels' power generation efficiency. In the case of SEU, the anomaly is a subsystem or mission failure.

and gamma rays, which can also contribute to surface charging and sensor damage (especially for remote sensing or optical hardware) (e.g., Bedingfield et al., 1996). Human decisions and activity can contribute to satellite anomalies as well, with such causes as operator error in commanding the satellite, electromagnetic interference ("jamming"—be it unintentional or intentional), and targeted attack by an adversary. Even mechanical spacecraft damage from micrometeoroids or space debris can cause issues that initially manifest as spacecraft anomalies, though such events are relatively rare. Two examples include micrometeoroid damage to a solar array during deployment on the international space station (NRC, 2011), and breaking of the Radio Plasma Imager 250-meter antennas on the IMAGE spacecraft, thought to be a result of micrometeoroid or debris impact (NASA, 2012a).

Here, we discuss several categories of satellite anomaly, caused either by the dynamic space environment or by human activity.

Total Ionizing Dosage (TID)

Persistent ionization because of solar electromagnetic radiation (the photoelectric effect) and bombardment by charged particles contributes to surface and internal charging. This buildup of charge can cause gradual degradation of electronics performance and solar panel efficiency over a period of years, but it can also result in electrostatic discharge that occurs instantaneously. When conducting spacecraft and mission design, engineers use an estimate of the TID that their spacecraft component can expect to endure during the desired mission lifetime. Hence, an estimate of the TID at a given altitude or region of space, and over a certain amount of time, is often used to design the radiation hardness level of a required satellite hardware component. This dosage is typically expressed as the amount of energy imparted to a material over a given amount of time, in units of radiation absorbed doses (Rads) (Si)[3] (e.g., Wertz and Larson, 1999, and references therein).

---

[3] 1 Rad (Si) = 100 ergs imparted per gram of silicon.

Electrostatic Discharge (ESD)

Electrostatic discharge (ESD) can occur when a strong-enough electric field builds up to cause an arc along the surface of the material, through the material, or between adjacent components. Such arcs can generate electromagnetic interference, which can cause anomalies in spacecraft operation (e.g., Robinson, 1989). ESD is usually the result of either surface charging or internal charging of the spacecraft. Hence, while ESD can result in an instantaneous and unexpected anomaly, it is also caused by a somewhat gradual physical process (the accumulation of charge over long periods of time, months to years). The rate of charge accumulation is not constant and depends on the plasma environment in which the satellite finds itself at any given time, as well as whether the satellite is in sunlight or eclipse. Also note that the threshold electrical potential required to produce an arc varies depending on the material and the spacing between conductors within the spacecraft components. The amount of internal or surface charge that may accumulate before ESD occurs may not be comparable between different satellites, especially those traversing different plasma populations within the magnetosphere.

The Anik E1 and Anik E2 satellite anomalies mentioned earlier are examples of problems caused by ESD. Both satellites were nearly identical in design and construction, and both were in geosynchronous orbit, exposed to very nearly the same plasma environment. The elevated fluence of high-energy electrons that led to the ESD in both satellites actually started nine days earlier, on January 11, meaning that the eventual ESD occurred after a nine-day buildup of charge on the spacecrafts' internal components. Note the two satellites' momentum wheel control systems failed within nine hours of one another (Lam et al., 2012).

Surface Charging

Surface charging is the buildup of electrical charge on the outer surfaces of the spacecraft that are directly exposed to the space environment. Ambient low- to mid-energy electrons (10–20 keV) impart their collective charge to the surface of the spacecraft. In addition, photons from solar illumination liberate electrons through the photoelectric effect, contributing to charging. The gradual buildup of charge can lead to different components of the spacecraft being at different electrical potentials, resulting in arcing that may damage some components. The rate of charge buildup is not constant, however, and there are times when the resulting ESD arcs are more likely to occur because of a sudden increase in potential, typically when the satellite moves from one plasma or illumination environment to another. For example, surface charging is a problem for high-inclination LEO satellites primarily because of passage through down-streaming plasma in the auroral zones during periods of high geomagnetic activity and, to a much lesser degree, cold plasma from the F-region of the ionosphere at lower latitudes (which peaks in free electron density at 300–400 km but extends to just above 1,000 km). At GEO, surface charging occurs intermittently during geomagnetic substorms, which inject high-energy electrons from the plasma sheet inward to GEO, especially between 2000 and 0800 magnetic local time (e.g., Mikaelian 2001; Romero and Levy, 1993; Wrenn and Sims, 2003; Gussenhoven et al., 1985). Also, anomalies stemming from ESD from surface charging may occur soon after the satellite leaves the Earth's shadow and becomes illuminated by the sun, since this exposure to solar photons provides a discontinuous "jump" in the spacecraft surface charge because of the photoelectric effect. The surface charge is more likely to reach an ESD threshold during one of these "jumps" than at other times during the gradual potential buildup (Fennell et al., 1985; NOAA, undated c).

Internal Charging

Internal charging is the buildup of electrical charge on the interior components of a spacecraft, typically leading to ESD. Also known as "deep dielectric charging," internal charging occurs when high-energy electrons (typically greater than ~10 keV) penetrate the spacecraft exterior and embed themselves into insulating materials, such as circuit boards and wiring insulation. If enough of an electrical potential builds up between internal components (~10 kilovolts) there could be ESD inside the circuitry of the spacecraft, leading to anomalies in operation. Most high-energy electrons are concentrated in the inner and outer electron

Van Allen belts, making high LEO orbits and GEO similarly suscep-
tible (e.g., Wrenn et al., 2002; Mikaelian 2001).

Recently, Lohmeyer and Cahoy (2013) analyzed failures of a
specific type of subsystem (solid-state power amplifiers, or SSPAs)
between 1996 and 2012, using data from the British telecommunications company Inmarsat, amounting to 26 anomalies over 16 years.
SSPAs amplify uplink signals received by the satellite and retransmit
downlink signals, making them critical components for satellite operation. The authors found that most of these anomalies occurred during
relatively quiet geomagnetic periods, but that a significant number of
SSPAs experienced a high-energy electron flux in the two weeks prior
to the anomaly. Also, the anomalies occurred in all local times, but
most occurred either in the midnight-to-dawn sector (pre-dawn) or the
local-noon-to-dusk sector (afternoon). It is difficult to build reasonable
statistics on such a small sample, but the authors inferred that most of
the SSPA anomalies were likely not caused by surface charging alone,
but rather by internal charging of components by high–energy electrons or a combined effect of surface and internal charging.

Single Event Effects (SEEs)

SEEs are anomalies caused not by a gradual buildup of charge over
time as with surface or internal charging, but by the impact of a single
high-energy charged particle into sensitive electronic components of a
satellite subsystem, this single event causing ionization and an anomaly. They typically occur because of high-energy (> 2 MeV) protons
and electrons striking memory devices in the spacecraft's electronics
systems, causing the spacecraft (or a subsystem) to halt operations,
either temporarily or permanently (e.g., Speich and Poppe, 2000).
SEEs include "bit flips" or SEUs, where a high-energy particle imparts
its charge to a solid-state memory device, causing errors in the system
software, which may or may not damage hardware and can potentially
be detected and repaired with error-detection-and-correction algorithms (EDACs) in the system software. One example of an EDAC is
triple-modular redundancy (TMR), in which three processors perform
the same calculations in parallel and then compare their answers. If
one processor's answers differ from those of the other two, the "cor-

rect" two would outvote the incorrect one, and the third processor system could be rebooted or otherwise corrected, and the subsystem in general continues to operate.[4] Other types of SEEs include single-event latchups (SELs), in which a subsystem hangs/crashes as a result of a high-energy particle impact. This causes the subsystem to draw excess current from the power supply, and the device must be turned off and then back on to be operable. Sometimes SEL can lead to destruction of the device if the excess drawn current is too high for the power supply. In this case, the SEE is referred to as single-event burnout (e.g., Wertz and Larson, 1999). Susceptibility to SEEs depends strongly on system design, and the risk is higher for satellites spending time in the Van Allen radiation belts or at GEO where there is a higher fluence of galactic cosmic rays and high-energy protons from Solar Proton Events (e.g., Mikaelian, 2001; Wertz and Larson, 1999; and references therein).

Faulty Hardware or Design

There is also the possibility that an anomaly may occur because of faulty hardware or software onboard the spacecraft. This cause is often coupled with the previously discussed space environment effects but may be more directly attributed to manufacturing or design error if the hardware fails when the space environment conditions do not exceed the design specifications, or when a part is inadequately shielded for the environment in which it is meant to be placed. Such hardware or design faults may be easier to identify if anomaly investigators are aware of whether other satellites using the same hardware are experiencing similar problems, or whether there are environmental causes that may be playing a role.

Operator Error

Operator errors are anomalies caused by humans incorrectly commanding the spacecraft in a way that causes abnormal or unexpected behavior. Examples of operator error include command error, causing

---

[4] Such a TMR system was used effectively on the space shuttle throughout its operational period and is a commonly advertised feature of single-board computers designed for use in satellites (e.g., Siceloff, 2010; Maxwell Technologies, 2012).

the satellite to take an action it was not designed to take, incorrect calculations of required thruster adjustments, reaction wheel rates, antenna pointing, power cycling, or failing to take action to "safe" the satellite when space environment conditions are known to be extremely hazardous, such as during a major geomagnetic storm. Some such storms can be predicted through observations of coronal mass ejections on the sun's surface, which are monitored by several spacecraft in Earth orbit (e.g., Solar Dynamics Observatory), in orbit around the sun itself (e.g., STEREO) or at the Earth-sun LaGrange point, about 1/100 astronomical unit closer to the sun than the Earth (e.g., SOHO, ACE).[5] Remote sensing observations of explosions on the sun's surface, if appropriately aligned, can provide more than a day's worth of warning before the solar plasma encounters the Earth's magnetosphere.

Potential Risks from Human Technological Actions
Anomalies may also be caused by other human technological actions, either accidentally or maliciously disrupting satellite operation.

Accidental disruptions primarily include harmful electromagnetic interference, or "unintentional jamming." This may be caused by adjacent satellites transmitting on similar frequencies with misaligned antennas, "cross-pole interference" stemming from misaligned uplink signal polarization, or terrestrial interference of downlink receivers. The result is that one satellite operator's transmission inadvertently overwhelms the receivers of another satellite, which is then unable to properly communicate with ground stations and has its mission or service disrupted (usually temporarily). Unintentional interference is a relatively common problem for communications satellites in GEO, where many satellites are concentrated and often use similar carrier frequencies. A vice president of the satellite operations firm Intelsat, which operates more than 50 communications satellites, has stated that the company typically deals with "thousands" of cases of unintentional interference every year (Shiga, 2007). A global industry organization known as the Satellite Interference Reduction Group (SIRG) has

---

5   One astronomical unit = 149,597,870.7 km, the mean distance between the Earth and the sun.

the goal of reducing this type of unintentional interference by raising awareness and encouraging standard practices and regulations. It has major satellite operators and analysis corporations among its participating members; e.g., Intelsat, Inmarsat, Eutelsat, Siemens, Aerospace Corporation, MITRE (SIRG, undated).

Another possible unintentional cause of satellite damage and anomalies is orbital debris, originating from rocket upper stages, nonfunctioning satellites, and—on rare occasion—spacecraft collisions. In 2009, the nonfunctioning Russian Cosmos 2251 satellite accidentally collided with the operational Iridium-33 spacecraft at a LEO altitude of 790 km (e.g., Iannotta and Malik, 2009). At a closing velocity of more than 24,000 km per hour (15,000 miles per hour) the collision destroyed the two spacecraft and created more than 1,875 additional pieces of debris along two orbital planes, as observed by the SSN. Though such collisions are exceedingly rare, the resulting debris can pose a threat to other satellites in similar orbits, and subsequent collisions could disable or destroy satellite subsystems, events that may initially appear as satellite anomalies. During the year 2010, NASA satellites conducted seven maneuvers to avoid accidental collision with orbital debris, four of which were specifically to avoid debris from the Cosmos/Iridium collision (NASA, 2011b).

There is also the potential that an adversary could intentionally cause anomalies in satellite operation. Possible means of doing so include cyberattacks on a satellite system's space or ground segment, radio jamming of command (uplink) or telemetry (downlink) transmissions, kinetic attacks using anti-satellite weapons, and the detonation of nuclear weapons at high altitude. Jamming of transmissions by intentionally bombarding a satellite's uplink transponder with high-power signals on a similar frequency is a capability available to a variety of groups worldwide. For example, the Falun Gong spiritual movement was able to jam a Chinese television satellite in this way in June 2002, interrupting television service to rural parts of China for eight days (Washington Post, 7/9/2002). Several nations have demonstrated the ability to destroy a satellite in orbit using an anti-satellite missile, including the United States (Eberhart, 1985), the former Soviet Union (Grego, 2012), and China (Broad and Sanger, 2007). And the detonation of nuclear weapons at high altitude during the tests of the early 1960s revealed an array of physical phenomena that heightened awareness of the need to "harden" satellite systems to withstand radiation and the space environment in the first place. The Ariel I, TRAAC, and Transit 4B satellites all suffered failures soon after the Starfish Prime 1.4 Mt nuclear weapon test at 400 km altitude on July 7, 1962, over Johnston Island. All three satellites became completely inoperable within 38 days, most likely because of solar panel degradation and ESD effects caused by the enhanced artificial radiation belts produced by the detonation (Hoerlin, 1976; Webb et al., 1995).

Any of these methods of attack, or cases of accidental harmful disruption, could first be detected via an anomalous report in a satellite's telemetry, or the loss of contact with the satellite altogether. In such a scenario, a centralized database of near-real-time satellite anomalies may help investigators ascertain the likely cause. For example, space weather events (causing increased surface charging, internal charging, SEEs, or degradation) may indiscriminately affect many satellites in a broad region of the magnetosphere. An intentional attack may affect only one or a few satellites in a more restricted region or controlled by a particular party. Accidental interference usually affects communication satellites with similar carrier frequencies over a confined angular region. These methods for distinguishing anomaly causes may not be universally reliable; for example, a high-altitude nuclear detonation (purposeful) or cloud of orbital debris (accidental) could affect many satellites indiscriminately over a broad region as well. Thus, successful diagnosis of anomalies would also depend on other sources of situational awareness such as space environment monitoring by NOAA and NASA, awareness of anti-satellite tests by other nations, SSN tracking of orbital debris, and commercial news of planned broadcast and communication activities by other companies. But an accessible centralized anomaly database could provide the raw record of the impact of these phenomena, enriching the ability to determine anomaly attribution.

Mitigation of Satellite Anomalies

Many satellite components are "hardened" to be able to withstand high fluxes of energetic charged particles. It is important to understand the context of spacecraft and mission design: The goal is to avoid anomalies as much as possible. Hence, while most anomaly investigations seek to determine the phenomenon that led to the anomaly, the satellite owner and manufacturer may be most interested in whether the anomaly is recurring and how it can be avoided in the future. Avoidance of future anomaly occurrences could be achieved through mitigation on the existing satellite mission or a new design attribute on a subsequent mission (usually increased shielding, different components, or different orbital architecture).

One example of mitigation on an existing mission is the suspension of observations by the Hubble Space Telescope when it is passing through the South Atlantic Anomaly (SAA), a region where the Earth's radiation belts dip closer to the Earth's surface because of a geographically fixed deviation in the Earth's magnetic field. The increased background of high-energy particles in this region at Low Earth Orbit means that Hubble's fine guidance sensors (targeting cameras) and other scientific instruments would be overwhelmed were it to conduct observations during that time (Space Telescope Science Institute, 2012).

With the notable exception of Hubble, this "safeing" of a satellite's instruments for passage through a particularly hazardous region of space or during a geomagnetically active period of time is a decision rarely made by satellite operators. For commercial satellite operators in particular, shutting down some of their asset's capabilities would result in lost functionality and revenue for a period of time. These operators may choose to forgo "safeing" their satellite to maintain operations, and therefore risk additional anomalies. This is a decision that varies from one operator (and satellite) to the next (O'Brien, 2012).

Science Satellites

Most spacecraft are unique in terms of radiation hardening, hardware components, and orbital trajectory (with some exceptions, including major constellations like the Global Positioning System in MEO and the Iridium communication constellation in LEO, or major telecommunications satellites built in series). As such, care must be taken when comparing satellite anomaly histories. For example, a space weather event may be powerful enough to overpower the shielding of one satellite but not of another. Examining the history of satellite anomalies among different satellites may not give a clear picture of where/when satellite hazards are phenomenologically highest. This may mislead a user interested in assessing the abundance of anomalies throughout the magnetosphere. To do that, one must compare events of only a particular satellite (or satellite design). Several empirical databases of anomalies exist from satellites that had the express mission of cataloging anomalies stemming from the dynamic space environment, including the Combined Release and Radiation Effects Satellite (CRRES), Spacecraft Charging at High Altitude (SCATHA), and Solar Anomalous and Magnetospheric Particle Explorer (SAMPEX) missions.

Figure 2.2 shows tallied anomalies from two science satellites (CRRES and SAMPEX) as a function of "L-shell." (The plot is adapted from O'Brien et al., 2009.) L-shell is a parameter corresponding to the distance from the center of the Earth in R_E at which a set of magnetic field lines connects to the geomagnetic equator. (For reference, LEO ranges from 1 to 1.3 R_E, MEO centers around 4.2 R_E, and GEO is at 6.6 R_E; Figure 2.3 shows a schematic to illustrate the concept.) A satellite's current L-shell essentially refers to which magnetic field line it is on at that moment. A satellite at L=4 might not be near the equator and might be much closer to the center of the Earth than 4 R_E, but might be saying it is somewhere along the dipolar field line that crosses the geomagnetic equator at 4 R_E. L-shell is often used to discuss regions within the inner magnetosphere because the space plasma environment there is largely organized along the Earth's magnetic field.

The top two plots in Fig 2.2 show anomalies caused by SEEs and internal charging (respectively) on the CRRES satellite, which was in geosynchronous transfer orbit with a perigee of 333 km (lower LEO) and apogee of 33,578 km (near GEO), and an inclination of 18.1 degrees. CRRES was launched on July 25, 1990, and its mission ended on October 12, 1991, when contact with the spacecraft was lost, presumably because of onboard battery failure. One of its primary missions was to expose a variety of satellite electronics to the natural space

Figure 2.2
Summed Anomalies on CRRES and SAMPEX Satellites

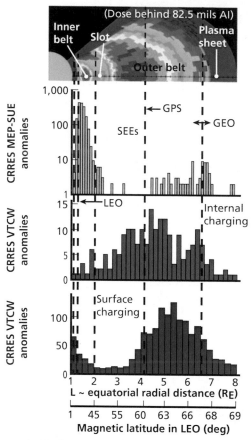

SOURCE: O'Brien et al., 2009. Image used with permission of Aerospace Corporation.
NOTE: Summed satellite anomalies caused by SEU and internal charging events on
the CRRES satellite (top and middle) and surface charging events on the SAMPEX
satellite (bottom). Shown as a function of L and magnetic latitude in LEO. Vertical
dashed lines show approximate locations of LEO, the slot region between the Van
Allen belts, MEO (GPS), and GEO. L corresponds to the equatorial distance from the
center of the Earth in RE, or the magnetic field line (at any latitude) that connects to
that radial distance at the geomagnetic equator.
RAND RR460-2.2

Figure 2.3
Illustration of L-shell Parameter

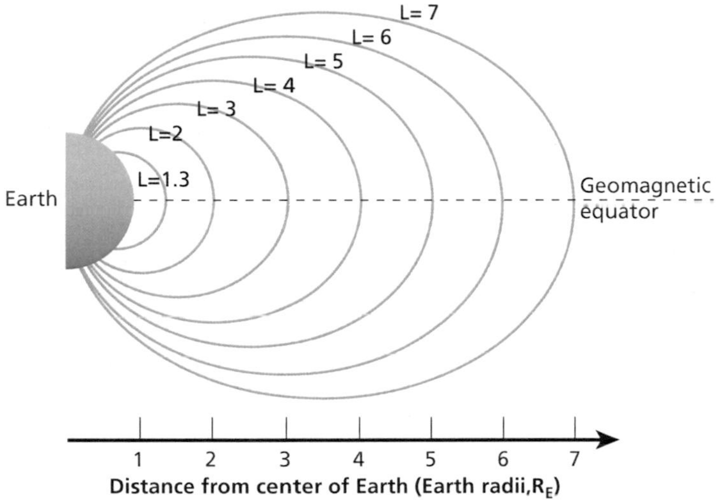

RAND *RR460-2.3*

environment and catalog the errors to determine the space environment conditions under which electronic failures were most frequent (Brautigam, 2002).

The bottom plot shows anomalies caused by surface charging on the SAMPEX satellite, whose mission was to measure magnetospheric energetic particles at LEO (perigee 550 km, apogee 675 km) with a high inclination of 82 degrees. SAMPEX operated from July 3, 1992, to June 30, 2004 (LWS Geospace Project Office, 2003).

Vertical dashed lines show the locations of LEO (upper and lower), the slot region between the radiation belts, MEO (GPS satellites), and GEO. Note that SEE anomalies (top plot) were highest as CRRES passed through the inner proton radiation belt, above 1,000 km altitude, but significant at GEO as well. Also note that internal charging anomalies (stemming from ESD) generally scaled with the density of the outer electron radiation belt, peaking in MEO (middle plot). This is consistent with our understanding of the outer belt, where high-energy electrons are able to penetrate the spacecraft surfaces and accumulate on internal components.

The surface charging anomalies (bottom plot) are not as well understood. They are from SAMPEX, which never flew beyond LEO. However, having such a high inclination, SAMPEX accessed high L-shells by traversing through high magnetic latitudes. The measurements on the bottom plot at L = 6.6 should not be misconstrued as measurement actually at GEO (as they are in the top two plots), but rather as measurements in LEO at high magnetic latitude. Note the surface charging anomalies on SAMPEX occurred either at low latitude (perhaps because of cold plasma and occasional passage through the SAA) or at high (but sub-auroral) latitude. Some studies have shown that the high L anomalies are well correlated with injections of high-energy electrons from the plasma sheet into the inner magnetosphere (Mazur and O'Brien, 2012), but this phenomenology has yet to be completely explained.

Anomaly Investigation Efforts

Currently, those tasked to investigate the cause of a satellite anomaly follow procedures that vary among satellite manufacturers and owners. Some companies that operate satellites they themselves have built (such as Intelsat) may also conduct their own anomaly investigations. Other satellite owners may not have the expertise to successfully conduct anomaly investigations and will defer to the spacecraft's manufacturer. At Orbital Sciences Corporation, for example, investigations begin when a customer who owns an Orbital-built satellite system reports an anomaly. Such perceived anomalies may be recurring resets of an onboard computer, or bit-flips that are more than just an annoyance. Investigators at Orbital attempt to trace the problem to the specific satellite subsystem involved, and then bring the subsystem engineers into the investigation. They obtain the subsystem data for six months prior to the investigation (as mentioned earlier, anomalies may be the result of a gradual buildup of charge over a long period of time), and often query NOAA's Space Weather Prediction Center to determine whether significantly abnormal particle fluxes occurred over that time period. Investigators also scrutinize the satellite design and particular hardware components, and whether any mitigation measures were being conducted or built into the systems (such as TMR or other EDACs). After risk-ranking the possible causes, they attempt to determine which causes are more or less likely, and they narrow the list to a specific root cause. Rarely is it a single root cause; more often it is a combination of multiple causes: for instance, insufficient shielding, hazardous space weather events, switching operation modes concurrent with an environmental change like a day-night transition, etc. (Noushkam, 2012). Often, these investigation efforts are conducted by Orbital without additional payment by the customer who owns the satellite, so Orbital investigators have articulated that any tool that could aid in the efficiency of such investigations would be a benefit to their business (Noushkam, 2012).

In the late 1980s and early 1990s, the Aerospace Corporation developed a software aid for the assessment of satellite anomalies caused by the space environment (e.g., Gorney and Koons, 1990). Their Spacecraft Environment Anomalies Expert System (SEAES) is a rule-based tool that helps an investigator determine the most likely cause of a particular satellite anomaly based on such user-input parameters as type of orbit, geomagnetic activity level (via geomagnetic indexes such as Kp), the "hardness" level of the circuits and components of the satellite system, etc. It essentially operates as a semi-automated software-based flow chart, taking the available information about the satellite anomaly and the known geospace conditions to estimate the likely category of anomaly (SEE, TID, ESD, etc.) and the likely cause. This expert system has recently been expanded with a particular emphasis in GEO satellites (O'Brien, 2009), making it highly useful for the myriad commercial communication satellites. There is also a version for use by human analysts, which simply provides visual flow charts without a software component (O'Brien et al., 2012). NOAA is also making efforts to develop a variant of this system called SEAES-RT (Real-time) (e.g., Darnel et al., 2012; Rodriguez et al., 2012). Tools such as this can be very useful in helping to determine the cause of a given anomaly, but their usefulness can be heightened when the inputs and results can be compared with those of many other satellites in a comprehensive database.

Future anomaly investigation methods may become even more automated in nature, to the point that anomalies are detected and categorized without any human involvement. One potentially viable option is to monitor various satellite telemetry streams with a ground system that detects anomalies automatically. Once the operator grants the system access to its telemetry stream (which could be encrypted), such a system could automatically analyze and determine which signals in the telemetry represent normal and abnormal events, and thus produce a running log of "anomalies." Years of previously archived telemetry could be used to "train" algorithms to identify anomalous occurrences, with human intervention and review needed only occasionally. Efforts have been made (for example, by AT&T Federal Systems) to evaluate this technique of using the "satellite as a sensor." The Aerospace Corporation has made efforts to develop a "Defensive Counterspace Test Bed for Spacecraft Attack/Anomaly Detection, Characterization, and Reporting" (Tschan, 2001). This type of machine–learning technology would be useful for identifying those anomalies caused by environmental effects, inadvertent interference, or even deliberate attack. The methodology essentially begins with choosing a cadence and then evaluating the state of the spacecraft at that regular time interval. Those investigating this technique warn against relying on "limit checking" (flagging as anomalous only those instances when a particular spacecraft system reports a value that is outside normal limits), since some abnormal behavior may occur within the limits. It is more thorough to look for any changes in behavior of the satellite subsystems, allowing a machine–learning algorithm to first determine what the "normal" behavior is, and then flag deviations from that normal behavior. Considerable care must be given to selecting the "training set" of data to feed to the algorithm, as the contents of that training set will determine what the algorithms regard as anomalous activity. Optimal selection of that training set is itself an area of active research (e.g., Tschan et al., 2012).

If a system is developed that can identify spacecraft anomalies with minimal effort by the operators themselves, it could be a major

improvement to space situational awareness, potentially allowing operators to avoid future satellite problems. It would certainly free up satellite operator personnel from manually searching through their spacecraft telemetry to do analysis on anomalies, instead enabling them to
review historical anomalies and determine trends that may be useful
in future spacecraft manufacture and design. Such an automated
system would, of course, also be useful in contributing to a centralized
anomaly database, extending its benefits to multiple owners/operators
(Tschan, 2012).

CHAPTER THREE

Anomaly Databases

Most satellite operators, be they civil, defense, or commercial, likely keep databases of their own satellite anomalies. However, detailed data for defense and commercial satellites are not typically shared publicly or among organizations. Also, these databases do not all conform to the same standard for which data is collected and cataloged, and in which format.

Here, we discuss examples of existing databases containing anomaly information from multiple satellite owners. We also discuss the concept of a potential future database that would be most useful for the operator community in diagnosing anomalies. The existing databases serve the purposes of particular members of the satellite operation community but tend to be either broadly available or comprehensive, but not both. A future database that is more comprehensive, broadly available, and centralized than the current offerings could improve the ability of the entire space community to efficiently identify and diagnose spacecraft problems.

Existing Anomaly Databases

Several anomaly databases have already been compiled, though usually they are not comprehensive and include only certain satellites from a certain range of time. Also, not all are broadly available in the community, and the anomaly information stored is not standardized from one database to the next. Here, we describe several existing anomaly databases to provide context on how a centralized anomaly database may add value.

NOAA NGDC Anomaly Database

The NOAA National Geophysical Data Center hosts an anomaly database that was managed by Dr. Joe Allen of NOAA (now retired). That database is publicly available and contains 5,811 anomaly records from as early as April 1983 and as late as December 1993 (Allen and Denig, 1993). It includes anomalies for commercial, civil government (NASA, NOAA), and DoD satellites. For each anomaly report, the database contains fields for anomaly date, start time and duration, uncertainty in time, spacecraft identification, orbit type (GEO, polar circular, elliptical, etc.), latitude and longitude of sub-orbit point when the anomaly began (and respective uncertainties), altitude, anomaly type (uncommanded status change, part failure, telemetry error, recoverable bit-flip, permanent chip damage, system shutdown, ESD measured, attitude control problem, or unknown), anomaly diagnosis (ESD from surface charging, electromagnetic pulse from internal charging, SEU, mission control problem, radio frequency interference, or unknown), the sun-vehicle-Earth angle, whether the satellite is spin or 3-axis stabilized, and a comment related to the anomaly. Availability of data for all the fields mentioned above varies from one record to the next. Many anomaly records are missing data, and some consist of only a time and a brief comment. Still, this database likely remains among the most comprehensive that are also publicly accessible. While detailed and extensive, this database has not been updated since 1993. Dr. Allen obtained anomaly information by actively maintaining contact with as many satellite operators as possible. After 1993, he was no longer able to maintain the database, and resources were not made available to continue the cataloging effort (Allen, 2012). One of the suggestions articulated at the 2012 NOAA Space Weather Workshop was to resurrect this database. There is interest within NOAA NGDC to do this, but funding and resources are currently limited for such a task (Green et al., 2012; Green, 2012).

Space News Digest

Peter C. Klanowski at the Space News Digest (SND) compiles a listing of satellite anomalies that is also publicly available and includes listings from 1993 up to the present day (Klanowski, 2012). For this database, each event is recorded with a date, the satellite identification, and a textual description of the anomaly providing whatever information is available. The textual description may include information on the subsystem, expected cause, specific time, or other pieces of information, but those data are not organized into table columns as the NGDC databases is. The database is updated regularly, but often the listings lack detail. As discussed previously, this is a result of the limited information commercial satellite companies and DoD satellite operators choose to disclose when one of their assets experiences on-orbit problems. Often, the only information provided is the Universal time (or perhaps simply the date) and a textual description of the anomaly (Klanowski, 2012; Allen, 2012). At least one scientific effort has been made to correlate the anomalies in the SND database with the geomagnetic activity index Kp, as well as charged particle densities at geosynchronous orbit (Choi et al., 2011). However, other groups have called into question whether the SND database is useful in anomaly attribution at all, owing mainly to the sparse amount of data available for each event (e.g., Mazur and O'Brien, 2012). For instance, only one-third of the SND events used in the Choi et al. study had universal times associated with them, making any conclusions drawn about the likely cause for most of the events imprecise at best.

Insurance Corporations

Companies in the business of insuring satellites against damage or failure have a strong financial interest in knowing the likelihood that a particular satellite design, component, or orbital trajectory is more or less prone to anomalies. The Atrium Space Insurance Corporation (ASIC) at Lloyd's of London Insurance Market insures approximately 190 satellites every year, representing about half of all commercial satellites. In order to estimate risk of failure, ASIC maintains a database with anomaly information for more than 922 satellites dating back to 1986 and up to present day (Wade, Hoffer, and Gubby, 2012; Hoffer,

2012a). The anomalies listed are mostly for satellites that are (or have been) insured by Atrium, though information on other satellites is added from time to time when possible, especially for other satellites carrying certain potentially problematic components that are also on an insured satellite. Atrium includes a clause in its contracts with satellite owners requiring that the owner report any failures or material changes to the health of the satellite when they occur. The client must also provide annual spacecraft health reports even if no significant failures occurred. Specialists at Atrium extract information from the client reports and enter it into Atrium's database. This database is quite detailed, including information on the type and manufacturer of most of the satellite subsystems, which allows Atrium to take note of any hardware that has experienced problems when used on previous missions.

The cause of a given anomaly does not typically affect whether the insurance company pays a claim or not. However, this detailed database helps Atrium to more accurately assess its risk when underwriting a particular mission (Hoffer, 2012b). While highly detailed and comprehensive of many satellites, the Atrium database is not available outside the company itself, as Atrium upholds a policy of not disclosing the details provided by its clients with anyone outside the corporation, including other clients. Hence, this database is not available for satellite owners or manufacturers as a reference during anomaly investigations.

Hypothetical Centralized Database

Existing databases of satellite anomalies are useful in that they provide historical and statistical data on the locations, subsystems, and conditions under which anomalies have occurred. These databases play a critical role in informing mission and hardware design specifications as a spacecraft mission is being developed. However, a centralized database with contributions from significantly more satellites in near-real-time could provide additional benefit, including the ability to utilize the current data to investigate current anomalies to determine their causes. Here, we discuss the motivation for developing a more centralized and accessible anomaly database, and what data might be included in it.

Articulation of Need from Satellite Community

Several groups have articulated interest in a centralized and standardized satellite anomaly database. In Mazur and O'Brien (2012), space scientists at the Aerospace Corporation stated:

> We (and others) have argued for an agency that would maintain adequate and open anomaly and abnormality lists . . . This information would enable robust statistical analyses of anomaly occurrence in order to develop statistical models and first-principles models of anomaly phenomena for improved satellite design and operations. [1]

Commercial satellite owners/operators at the 2012 NOAA Space Weather Workshop, hosted by NOAA's National Geophysical Data Center and Space Weather Prediction Center, also made a strong request for such an agency. At a special one-day Satellite Anomaly Mitigation Stakeholder's meeting, several commercial companies indicated that they could more quickly diagnose anomalies as being related to the space environment or other causes if they were made aware of whether other satellites were experiencing similar problems (Green et al., 2012).

Suggestions for organizations to manage such information typically focus on noncommercial, nonprofit entities with no vested financial interest in learning about the health of other organizations' satellites. Essentially, what is needed is an unbiased third party. Suggestions by members of the satellite community for organizations meeting this requirement have included the Aerospace Corporation (an FFRDC) and NOAA (Green, 2012; Noushkam, 2012).

We note that insurance corporations such as ASIC have already compiled a significant amount of data for commercial satellite anomalies, and have incentive to continue to do so as it helps them to calculate

---

[1] In the referenced paper, "anomaly" and "abnormality" were used interchangeably. For more on statistical models, see, e.g., O'Brien (2009). For more on first-principles models, see, e.g., Davis et al. (2008).

the risk of insuring satellites with certain components and orbital characteristics. They also have incentive to keep the data secret to maintain their customer base of operators who value the discretion. Though they typically have strict policies against sharing the data outside the company, even with their clients, the fact that they have already aggregated the data in a consistent format means that they would only have to anonymize it and find a way to make it available to those investigating anomalies without compromising the privacy of their clients. For example, the insurance company (or a consortium of insurance companies) could offer discounts or benefits to those clients willing to allow the insurers to make their data available in an anonymized form. An "opt-in" policy like this may not interest all clients, but it would be a positive step toward useful sharing of a comprehensive and up-to-date database managed by a motivated, trusted, and international third party.

Necessary Data

The Aerospace Corporation produced a brief technical report with recommendations on the minimal information required for an anomaly database to be useful for correlating anomalies with space environment activity (O'Brien et al., 2011). Note that, even if a non-space-environment cause is suspected, it is still important to consider space environment effects as a contributing factor. Table 3.1 shows a list of required and recommended contents for a useful satellite anomaly database, slightly modified from the original list by O'Brien et al. (2011) to include two additional suggested items that may be useful. Columns show, respectively, whether the listed item is considered a "requirement" by O'Brien et al., whether the listed item reveals the identity of the satellite by itself, and whether the listed item could reveal the identity of the satellite when combined with other listed items. An item is marked "Depends" when the information could potentially be used to reveal the identity of the satellite, depending on which specific other information is listed in the database.

O'Brien et al. (2011) based these recommendations on the various databases and lists of satellite anomalies they analyzed while developing the SEAES (O'Brien et al., 2009). They also assert that items

Table 3.1
Suggested Contents of an Anomaly Database

Description*	"Required"*	Reveals Identity (alone)	Reveals Identity (combined)
1. Date and Universal Time of anomaly	x	No	Yes
2. Fully specified spacecraft location during anomaly	x	Yes	Yes
3. Velocity or orbital elements at time of the anomaly	x	Yes	Yes
4. L-shell at time of anomaly		No	Depends
5. Magnetic Local Time of vehicle during anomaly		No	Depends
6. Eclipse state of the vehicle (full, penumbra, partial, none)		No	Depends
7. Vector to sun in spacecraft coordinates		No	Depends
8. Velocity vector of spacecraft in spacecraft coordinates		No	No
9. Initial guess at type of anomaly (SEU, discharge, TID)		No	No
10. Estimated confidence of that guess		No	No
11. Anomaly category (e.g., affected subsystem or type of disruption)		No	Depends
12. Vehicle identity (possibly anonymized)		Yes	Yes
13. Notes on recent operational states or changes (e.g., recent commands, attitude schemes)		Depends	Depends

NOTE: * Modified from O'Brien et al. (2011).

1-3 are "required" for an anomaly database to be useful in diagnosing statistical relationships between anomalies and potential causes. Many other databases and anomaly lists have contained some of the data listed in Table 3.1, but at the time of this writing there has been no widely accepted standardized format.

Discussions with other experts in satellite anomaly investigation corroborate the data types recommended above, though the order of importance may vary in the opinions of different satellite operators and investigators. For example, some suggest that the most useful pieces of information for an investigator would be the general orbital region

(GEO, LEO, low vs. high inclination, highly eccentric orbit, etc.) followed by the satellite subsystem (which is listed at item 11 in Table 3.1) (Noushkam, 2012). Mazur and O'Brien (2012) suggested that an agency be created to maintain open satellite anomaly lists that contain, at a minimum, the subject vehicle, the date and time of the event, the 3-D location of the vehicle at the time of the event, the 3-D velocity or complete orbital elements for the vehicle at the time of the event, the affected subsystem, the suspected type of anomaly, and the level of confidence in that assessment.

There is a general consensus that the time and position of the satellite are critical, with many other data points potentially helpful in contributing to the anomaly investigation and diagnosis process (e.g., Green, 2012; Allen, 2012; Noushkam, 2012). The spatial and time scales associated with certain types of anomalies may also help in their diagnosis. As noted earlier, because of the nature of plasma injection into the inner magnetosphere during a geomagnetic storm, anomalies associated with geomagnetic activity often occur in geosynchronous satellites in the predawn sector, and the main phase of a geomagnetic storm tends to last from two to eight hours. As such, the precise position of the satellite may not be necessary if the database at least reveals whether, for example, other satellites in the predawn sector experienced anomalies within the past six hours.

A centralized database would be ultimately useful for anomaly diagnosis and future satellite design and development efforts. However, depending on the rate at which participating satellite owners contribute to the database, and the rate at which the managing entity updates it, the database may be useful in "near real-time" to allow assessment of ongoing problems. For example, if a satellite were experiencing communications problems originally caused by unintentional (or intentional) jamming from other satellites or terrestrial sources, rapid reporting of these anomalies and sharing the information via a database may lead to multiple owners realizing the problem is limited to a particular region of the GEO belt, and diagnosing jamming as the likely cause. To be useful for this type of real-time situational awareness, the database would likely need to be updated hourly, since telecommunications companies have strong incentives to restore communications in as little time as possible. Hourly updates of the database may be unrealistic, unless both an automated updating and management system runs the data collection and distribution, and participants are diligent about sending in their updates to the central database even while they are dealing with an ongoing communications outage with their on-orbit assets. Daily or weekly updates, however, are more achievable and would still aid in post-anomaly diagnosis.

We also note that the SND and ASIC databases largely use textual descriptions to document all available information for the anomaly, whereas the NGDC database used a table with particular columns into which data was entered if available, and left blank if not. To enable quick and effective analysis of anomaly data, information must be encoded in searchable/sortable fields, not simply listed as free text. Hence, it is important that all users agree on what fields should be used and what range of values can be entered; this will ensure the database is useful for analysis. Additional textual data can always be stored in a "comments" column.

Knowledge of the design details of the satellites experiencing anomalies is important when attempting to diagnose them. This is why most anomaly investigations are carried out by the organization that built the satellite. When checking a database for other satellites that experienced similar anomalies, it would be useful to be aware of whether those other satellites had similar components or hardening levels. One example of data that would be useful to store in a free-text "comments" column would be any information on particular components or design specifications of the satellite and its subsystems. In the case of the ASIC database mentioned earlier, the insurance company collects a great deal of design and component information so that it can better estimate risk with regard to components that have had previous anomalies or failures on other satellites.

In Table 3.1, the "Reveals Identity (alone)" column indicates whether the particular category of information could be used to reveal the identity of the satellite or operator, should that information alone be available publicly. The "Reveals Identity (combined)" column indicates whether the particular category of information could be used to reveal the satellite identity when combined with other information listed in the database. For example, the Universal Time of the anomaly (item 1), taken alone, would not reveal a satellite's identity. But when that is combined with the position of the satellite, any other user may be able to use the combined information to identify the satellite experiencing the anomaly. This is relevant because some contributors (especially commercial and DoD) may wish to contribute to and benefit from the database but would want to keep secret the identity of the satellite experiencing problems. As shown in the table, it is very difficult to make a shared and broadly available database that is useful yet maintains the anonymity of the satellites. We will discuss this further in Chapter Four.

Meeting the Security Requirements of Contributors

Commercial and DoD operators are often unwilling or unable to share the precise position and velocity of their satellite at a given time (their state vector), as such information could be used to determine the satellite's orbital trajectory and identity, information the contributor may hold as proprietary or classified. Commercial satellite owners have articulated that they may prefer not to reveal the identity of the spacecraft experiencing anomalies, as it could affect customer confidence. Our discussions with subject matter experts have indicated that these privacy and security concerns likely pose the most significant obstacle to the creation of a centralized and shared anomaly database (Noushkam, 2012; O'Brien, 2012).

We identify two fundamentally different privacy concerns related to developing a database of anomalies:

1. privacy concerns related to incorporating sensitive data into a trusted database
2. privacy concerns related to users' access to the secure database.

The first issue addresses privacy concerns in the creation of any database of anomalies. If an aggregate database of anomalies were created, who would store and maintain it? If an agency (e.g., NOAA, an FFRDC, or a consortium of insurers) were to maintain such a database, any contributing satellite operator must inherently trust the agency with its private anomaly data. Even if operators are willing to trust the agency, the agency itself may not be willing to assume the security responsibilities that are associated with that trust.

The second issue addresses a possible loss of privacy stemming from anomaly properties stored within the database. Even if a trusted third party were to host and maintain a database of anomalies, users could violate the privacy of the contributing operators simply by accessing the database. For example, O'Brien et al. (2011) suggested that any useful database would catalog the precise time, location, and velocity at the time of the anomaly. Any user with access to the database could likely identify the satellite owner from this information alone. This indicates a trade-off between utility of the database and privacy of the contributing operators. As the database grows richer, the individual assets of contributing operators can be more easily identified from simple database queries.

Both of these issues can be partially addressed with modern cryptography. The first privacy concern—that of finding a trusted party to host such a database—can be solved using techniques in secure MPC. The second privacy concern—that of information leakage through querying the database—can be addressed using tools from differential privacy (Dwork, 2006). These solutions are completely independent and can be implemented individually or in tandem.

A Secure Multiparty Computation Solution

Secure MPC is a cryptographic tool that allows a collection of participants to compute any function of their private inputs while provably maintaining the privacy of each individual's input. The theoretical framework for secure MPC was laid in the 1980s in a series of works by academic cryptographers (Yao, 1982; Yao, 1986; Goldreich, 1987; Ben-Or, 1988). Since that time, secure MPC has been an active area of cryptographic research. Within the cryptographic community, the applications of secure MPC to privacy-preserving data-mining are well known, and Lindell and Pinkas provide a survey of potential applications (2009). Intuitively, secure MPC protocols can be viewed as a tool to allow a collection of individuals to achieve anything that could be achieved in the presence of a trusted third party, without the need for mutual trust. In particular, a transparent and provably secure cryptographic algorithm replaces the trusted party. Thus, secure MPC can be used to address the first privacy concern noted above, that of hosting a centralized database of sensitive anomaly information. On the other hand, secure MPC does not provide more privacy than a trusted third party could provide. In particular, secure MPC does not address the second privacy concern, that of privacy violations that arise from responding to queries. Take, for example, a database containing the time and location of all anomalies: If a user queried the database and found an anomaly reported in a region of space and time where only one operator was known to have satellites, the user could deduce the identity of the operator reporting the anomaly. This type of privacy violation can occur whether the database is maintained by a completely trusted party or virtualized using a secure MPC protocol. Privacy violations that occur from *responding* to queries are not addressed by secure MPC, and in the next section we discuss using differential privacy to combat leakage resulting from query responses. In this section, we outline how secure MPC could be used to create a virtualized anomaly database for the satellite community without the need to find a trusted host for the community's sensitive information.

To apply secure MPC in the setting of satellite anomalies, each participant (satellite operator) maintains a private catalog of anomalies affecting the operator's own assets. In a secure MPC-based approach, no aggregate database is created, and thus no trusted party is needed to store and maintain it. When an operator wishes to learn statistics about the aggregate data set (e.g., how many anomalies were reported during a specific time interval in a specific region), the operators engage in a secure MPC protocol to compute the desired result. The operator initiating the query uses it as his private input to the MPC protocol, and each other operator's private input to the MPC protocol is its own anomaly catalog. The security of the MPC protocol then guarantees that nothing is revealed about each operator's catalog beyond what is revealed by the results of the query.[1] The security of the protocol also guarantees that the query itself remains private. Maintaining the privacy of the query is important, as the query itself may reveal information about the party making the query. For example, a query like, "how many anomalies were reported in region X?" might indicate that the individual making the query had experienced such an anomaly.

Utilizing secure MPC to answer user queries completely eliminates the need for a trusted party—no aggregate database is actually created, and no aggregate database needs to be stored or maintained. Instead, individual users can engage in a secure MPC protocol that can be viewed as querying a virtual database. Since each user maintains a private anomaly catalog and never shares it, no mutual trust is needed among the operators, and no trusted outside party is needed to facilitate collaboration. The only communication between the operators is within the framework of the MPC protocol, and thus the privacy of each operator's anomaly catalog is guaranteed by the security of the protocol. The protocol also ensures that each operator's queries remain completely private.

While such an MPC-based approach completely eliminates the problem of trusting an outside party with sensitive data, it does have drawbacks. The primary one is that it requires interaction among the contributors. Each time a user wishes to learn aggregate anomaly information, all contributors must engage in a secure MPC protocol. In particular, making a query to the database requires interaction among all the contributing operators. Thus, each contributor must ensure that its private server is online and available to participate in the computation of each query's response. Maintaining a networked server may not be a prohibitive cost, but it is a cost that is not present in the trusted third-party model, where each operator can hand its private anomaly catalog to the trusted party and go offline.

---

[1]   While query results may leak sensitive information, secure MPC does not address this type of privacy violation. To prevent this type of leakage, different tools are needed, and a differential privacy solution to combat leakage that would naturally result from query responses is discussed in the next section.

While secure MPC is widely recognized as a great theoretical breakthrough within the cryptographic community, the inefficiency of existing protocols has prevented its widespread use in practical situations. In fact, while any function can be computed securely using an MPC protocol, such a secure computation can be orders of magnitude slower than the corresponding insecure calculation. For many years, any type of secure computation was seen as impractical. However, the rapid pace of development on the underlying cryptographic algorithms has pushed many previously inefficient solutions into the realm of the practical. In this situation of securely aggregating anomalies, the database queries envisioned are fairly simple (e.g., counting the number of anomalies satisfying certain criteria), and the resulting secure computation can likely be made to be extremely efficient.

Secure MPC protocols provide a means for converting a publicly known function on private inputs into a protocol that securely computes the result of the public function, without revealing each participant's private input. In this case, the function takes a list of anomalies from each operator, and a database query (e.g., "how many anomalies occurred in a given region in a given time-frame?") and outputs the anomalies that satisfy the criteria specified by the query.[2] At a high level, most secure MPC protocols work by first transforming the public function into a circuit that computes the same functionality. Using the (public) circuit representation of the function, the participants in the protocol work to securely evaluate the circuit one gate at a time. Secure MPC protocols are often inefficient for two reasons: First, the circuit representation of a function may be extremely large; and second, securely evaluating each individual gate of the circuit requires performing cryptographic operations which can themselves be fairly computationally intensive. Improving the computational efficiency of secure MPC protocols is currently an active research area within the cryptographic community.

---

[2] This is just one example of a simple functionality that could be computed using secure MPC. Secure MPC protocols could be used to securely emulate any other database functionality. Thus, instead of returning the list of matching anomalies, it could return only a count of the number of anomalies that matched, or the average number of matches per hour, or the number of distinct satellites that experienced anomalies, etc.

Closely related to secure MPC is the notion of private informa-tion retrieval (PIR) (Ostrovsky and Skeith, 2007). PIR is a crypto-graphic tool to allow an individual to query a database while keeping the query itself private from the database. Most PIR protocols work by allowing a client to create an encrypted query in such a way that the database operator can run this encrypted query on the database to generate an encrypted response. The privacy ensures that the encrypted database response cannot be decrypted by the database operator (that would leak information about the query) but only by the client who generated the encrypted query. For a survey of existing PIR protocols see the work of Ostrovsky and Skeith (2007).

Although efficient and secure PIR protocols exist, the question of building a secure, collaborative anomaly database does not fit well within the PIR framework for many reasons. First, building a commu-nity anomaly database is a question of how individual operators can combine their private catalogs in a way that is both privacy-preserving and beneficial to the community, and not simply a question of how an existing database can be accessed privately. Second, most PIR proto-cols are not designed to limit the amount of information obtained by the database client. Third, PIR protocols are not easily combined with tools like differential privacy (described in the next section) that limit the leakage from query responses. Thus, traditional PIR protocols will likely not be relevant in this scenario, and more general MPC protocols are needed.

A Differential Privacy Solution

Once an aggregate anomaly database is created (either under the con-trol of a trusted party or virtualized through a secure MPC), the ques-tion remains whether simply providing access to the database can violate the privacy of contributors. For example, as the entries in the database increase in specificity, it is more likely that individual assets can be uniquely identified by publicly available characteristics in the database. On the other hand, the value of the database is diminished if fewer elements of the anomaly could be revealed to the community.

This highlights a fundamental trade-off between the utility of the database and the privacy of the contributors. To address these concerns, various techniques can be adopted to further anonymize the database records while minimally impacting the utility of the database as a tool in identifying patterns of anomalies.

For example, it may be that simply revealing the position and velocity of a satellite during the recorded anomaly still divulges too much information about the identity of the specific satellite. To combat this, instead of giving position and velocity of the satellite that encountered the anomaly, the database could instead record the satellite's local time or L-shell. Recall that L-shell is a parameter corresponding to the distance from the center of the Earth (in R_E) at which a set of magnetic field lines crosses the geomagnetic equator. A given L-shell value prescribes the satellite's location along a particular shell of magnetic field strength, which is important in understanding what part of the magnetosphere it is in—and, therefore, which plasma populations it is exposed to. Thus, revealing the L-shell at the time of anomaly may be almost as useful as revealing the exact position. But without the magnetic latitude, the exact location of the satellite is still partially hidden: That is, since L-shell values without a magnetic latitude don't imply a particular distance from the Earth, the precise position and satellite orbit-type may still be concealed, while the information useful for anomaly investigation is shared.

While such anonymizing techniques may prove useful, it can be difficult to assess their impact on the utility of the database and the privacy of the contributors. For example, exactly how much information is leaked by revealing a satellite's L-shell? Is this below an acceptable threshold? How much utility is gained by learning only the L-shell where an anomaly occurred? Answering these questions may be difficult, or impossible, in the case of anonymizing via L-shells. Examples of ad hoc anonymization techniques that have failed to achieve the desired level of privacy can be found in the survey of Heffetz and Ligett (2013).

While analyzing the leakage of various ad hoc anonymizing techniques is often difficult or impossible, approaching anonymization in a statistically rigorous way allows the leakage to be quantified and bounded, and privacy can be ensured.

One successful framework for developing and studying anonymizing techniques is the notion of *differential privacy* (Dwork, 2006; Dwork, 2008), which provides very general and provably secure anonymizing techniques—and "differentially private" anonymizing provides the strongest possible guarantees of security for database contributors.

Given a database of sensitive information, a mechanism for accessing these data is called differentially private if the impact on any individual's privacy is the same whether or not his data is included in the database. In the satellite scenario, this means that a mechanism implementing an anomaly database would be differentially private if the amount of information leaked about any operator's assets would be essentially the same *whether or not his assets were included in the database.* This is a very strong privacy guarantee and can be applied very generally.

Returning to the example of anonymizing via L-shells, it can be very difficult to estimate just how much privacy is gained (and utility is lost) by revealing only the L-shell of the satellite experiencing the anomaly. On the other hand, by using a formal differentially private mechanism, the exact amount of privacy gained (and utility lost) can be quantified and analyzed.

Differentially private systems work by adding small perturbations to the values in the database. The difficulty lies in finding perturbations that are small enough that they do not affect aggregate statistics, but are large enough that they can hide individual records in the database. To create useful and powerful differentially private mechanisms, it is important to allow the perturbations to vary. Thus, to achieve differential privacy, a curator sits between the database and the client. The curator moderates the queries posed by the client and the subsequent responses from the database to ensure differential privacy for the contributors to the database.[3] Essentially, the curator's role is to

---

[3]   This is the notion of interactive differential privacy. A separate notion of non-interactive differential privacy exists, where the database is processed once, then released to the public without any further intervention by a curator. This processing must be such that it effectively protects the privacy of the contributors while still maintaining the utility of the database.

add perturbations to the database responses in such a way that privacy is ensured for each record in the database. Designing a differentially private mechanism, then, amounts to finding distributions of perturbations that have minimal effect on the statistics being computed and provide mathematically provable privacy guarantees for the underlying data. Differentially private database mechanisms exist, and efficient solutions strive to achieve the maximum privacy protection while incurring the minimal loss in utility. An overview of the mathematics of differential privacy can be found in the survey of Heffetz and Ligett (2013).

While differentially private mechanisms exist in general, their efficiency depends on what types of queries are being made. In the setting of anomalies, the class of useful queries is extremely simple, making it especially suitable for a differentially private mechanism. The most useful queries are likely "count" queries; e.g., counting how many anomalies occurred in a specific time frame, and specific region. Differentially private mechanisms for count queries can be made exceptionally efficient (Dwork et al., 2006).

It is important to note that differentially private mechanisms can be employed whether operators use an approach based on a trusted third party or a secure MPC. If a trusted party (e.g., NOAA, an FFRDC) is employed to manage the aggregate database, it can also serve as the curator, modifying responses to database queries in accordance with the specifications of the differential privacy mechanism. For example, if several satellites all in the same region of local time, orbit, or altitude experience an anomaly, the trusted party/curator may alert only those owners whose satellites are in the same region. Alternately, if the database contains information about satellite design factors or components with a history of anomalies under certain conditions (like the ASIC database mentioned earlier), especially those that are common to multiple satellites, the users could query whether other satellites with similar components experienced anomalies recently or in the past under similar conditions. If the operators replace the trusted party with a secure MPC, the function-

---

We will focus on the interactive setting, however, because the interactive solutions for differential privacy are significantly more powerful than their non-interactive counterparts.

ality of the curator can be easily rolled into the secure computation itself, thus ensuring that users only ever see the curated output of their queries.

While using secure MPC to implement the differentially private mechanism may be inefficient, full-blown MPC is not needed for some simple types of queries. For example, the DJoin system allows users to calculate differentially private database "join" queries without the need of a trusted curator using an efficient lightweight MPC protocol (Narayan 2012). Using a tool like DJoin, satellite operators could implement simple count queries (e.g., "How many anomalies of a certain type occurred in a certain region?") in a differentially private way without the need for a trusted party to host the database, and without the computational overhead of a full secure MPC protocol.

Cryptographic methods such as secure MPC protocols and differentially private mechanisms may fall within the purview of DARPA's PROCEED program. These techniques could be applied and tailored to the problem of securely sharing satellite anomaly information as a use case for PROCEED, contributing to improved anomaly diagnosis systems that would benefit the broad community of satellite operators.

Remaining Privacy Concerns

Specificity of Anomaly Records

Regardless of whether operators aggregate their data through a trusted party or through a secure MPC protocol, individual contributors can decide which pieces of anomaly data they will share at any given time. For example, an operator may prefer not to reveal what exact component experienced the anomaly to avoid implying a particular manufacturer is at fault before more information is available. Perhaps they can just reveal the subsystem or functionality that was affected (e.g., navigation or communications). If a differentially private mechanism is in place, then users can always contribute their most accurate information without a privacy risk.

Inclusion of Classified Sources

While an aggregate anomaly database would benefit all satellite operators, civil government agencies and commercial satellite operators would be the most likely contributors. Satellite operators in DoD that manage classified assets may be less likely to contribute. Currently, the Aerospace Corporation manages classified anomaly databases for some DoD satellites (O'Brien, 2012). It is likely these classified databases would remain the preferable method for cataloging DoD anomalies, as sharing some of the information outside DoD could be considered an unacceptable risk to national security. Indeed, it is possible that DoD might not accept any third party or commercial entity outside of the United States government as the steward of a centralized database including comprehensive information on DoD satellites. That said, whether agencies like the DoD choose to contribute information or not, the information contributed by the community could still be used to benefit DoD by augmenting and comparing with any internal classified databases that are being maintained. Note, however, that there is precedence for operators of DoD satellites contributing some (limited) anomaly information to the NGDC database—only those data the DoD is comfortable releasing publicly. These contributed data may be as simple as only a date or time but may still be more useful to other operators than a lack of information. The DoD and its satellite operators would have to use their discretion to determine what anomaly information they are willing to contribute to a centralized database.

Public vs. Private Output

Any centralized anomaly database could also have different tiers of access, whether it is implemented by a trusted party or a secure MPC algorithm. For example, only a small amount of generalized statistical information could be made available to the general public ("approximately ten satellites experienced upsets within the past day in the pre-dawn sector"), but more detailed information could be available to agencies with a higher clearance. A similar tiered structure could be used to distinguish access between domestic and international users, which may be important to both DoD operators and commercial users for compliance with the International Traffic in Arms Regulation. A balance must be struck between how much data would be useful and how much data the majority of participants would be willing to share. Exactly which information is provided to which users is beyond the scope of this report, but this serves to highlight the versatility of a secure MPC protocol: Any tiered access structure that could be implemented with a trusted party could also be implemented using secure MPC, without the need of a trusted party.

CHAPTER FIVE

Observations and Recommendations

Having conducted a literature review, discussions with subject matter experts, and an overview of potentially useful encryption strategies, we arrive at the following observations and recommendations.

- A centralized and standardized satellite anomaly database is recognized by subject matter experts from NOAA, the Aerospace Corporation, and numerous commercial companies as a potentially valuable resource for the satellite operator community. Such a database would aid in anomaly investigations, thus reducing costs and increasing efficiency. As a side benefit, it could also contribute to the scientific understanding of the real-world impacts of the near-Earth space environment, consistent with the scientific goals articulated by the National Science Foundation's Geospace Environment Modeling (GEM) program (2012), and NASA's Living With a Star (LWS) program (2012).
- A single, centralized database could offer advantages over multiple, smaller databases. Multiple smaller databases already exist, as described in Chapter Three. However, they tend to be either broadly available but incomplete (e.g., the NOAA NGDC and SND databases) or highly detailed but not broadly available (e.g., the ASIC database). Individual satellite operators often maintain their own internal anomaly databases, a practice that would be useful for DoD as well. But these multiple databases vary in accuracy and content. A centralized database could reduce the duplication of effort involved in maintaining multiple databases with

the same (or similar) information, while providing data that is both detailed and broadly available.

- The development of a centralized satellite anomaly database that would be useful to the broad satellite community is hindered by concerns about sharing proprietary information, as well as the lack of available resources to develop and maintain such a database. Concerns over sharing of proprietary information are perhaps the most significant obstacle for companies in the commercial sector. The lack of resources for development and maintenance is a problem in the civil or defense sector of government, which would likely include organizations that could serve as trusted third parties, and for those that could develop encryption technologies that could obviate the need for a trusted third party.

- Cryptographic techniques such as secure MPC and differential privacy may help overcome inhibitions of commercial satellite operators to share anomaly information, thus contributing to greater benefit throughout the satellite operator community. The proposed methods discussed in this report are merely overviews of relevant concepts, but they show promise and could be further developed and applied to contribute to a comprehensive solution. This is a potential use case for DARPA's PROCEED program, which seeks to develop methods that enable computing with encrypted data without first decrypting it.

- Automated "satellite as a sensor" methods for identifying and cataloging anomalies may significantly reduce the workload of those investigating satellite anomalies. Such systems can enhance both cataloging and categorization efforts and improve space situational awareness. They can also be used as an alternative to data sharing to help diagnose anomalies from the telemetry of a single satellite.

In summary, additional research into applying the methods discussed to satellite anomaly identification and secure sharing could enable enhanced and efficient anomaly diagnosis by the broader satellite operator community. Ultimately, for such efforts to succeed, the commercial satellite operator community must be convinced that the sharing mechanism meets their privacy standards and that the benefit of sharing is worth the effort and perceived risk. It is likely that commercial and DoD operators would provide only limited information to such a database at their own discretion, but given that no publicly available, trusted, and updated database currently exists, this would still be an evolutionary improvement. Another positive step would be allotment of resources to those third-party entities that would be trusted to serve as stewards of the database (i.e., federal agencies or FFRDCs), or innovative business plans that would enable stewards of existing private databases (such as insurance corporations) to facilitate sharing of anomaly information via incentives for clients willing to participate, or aggregated publication of anomaly data that does not reveal identity.

References

Allen, Joe, and William F. Denig, *Satellite Anomalies*, database, National Oceanic and Atmospheric Administration, National Geophysical Data Center, 1993. As of December 6, 2012:
http://www.ngdc.noaa.gov/stp/satellite/anomaly/satelliteanomaly.html

Allen, Joe, physicist (retired), NOAA National Geophysical Data Center, telephone and e-mail communication with the author, August 31, 2012.

Bedingfield, K. L., R. D. Leach, and M. B. Alexander, ed., *Spacecraft System Failures and Anomalies Attributed to the Natural Space Environment*, National Aeronautics and Space Administration, NASA Reference Publication 1390, August 1996. As of October 3, 2013:
http://maelabs.ucsd.edu/mae155/classes/wi_05/space%20envt_nasa%20rp1390.pdf

Ben-Or, Michael, Goldwasser, Shafi Wigderson, Avi, "Completeness Theorems for Non-Cryptographic Fault-Tolerant Distributed Computation," *STOC '88*, New York: ACM, 1988, pp. 1 10. As of October 3, 2013:
http://doi.acm.org/10.1145/62212.62213

Brautigam, D. H., "CRRES in Review: Space Weather and Its Effects on Technology," *Journal of Atmospheric and Solar-Terrestrial Physics*, Vol. 64, No. 16, 2002, pp. 1709–1721. As of December 6, 2012:
http://www.sciencedirect.com/science/article/pii/S1364682602001219

Broad, William J., and David E. Sanger, "Flexing Muscle, China Destroys Satellite in Test," *New York Times*, January 19, 2007, 2007. As of December 12, 2013:
http://www.nytimes.com/2007/01/19/world/asia/19china.html?pagewanted=all&_r=0

Choi, Ho-Sung, Jaejin Lee, Kyung-Suk Cho, Young-Sil Kwak, Il-Hyun Cho, Young-Deuk Park, Yeon-Han Kim, Daniel N. Baker, Geoffrey D. Reeves, and Dong-Kyu Lee, "Analysis of GEO Spacecraft Anomalies: Space Weather Relationships," *Space Weather*, Vol. 9, No. 6, 2011, p. S06001. As of December 6, 2012:
http://dx.doi.org/10.1029/2010SW000597

Darnel, J., et al., "Implementation of Space Environmental Anomalies Expert System Real Time," poster presented at NOAA Space Weather Workshop, Boulder, Colo., National Oceanographic and Atmospheric Administration, April 23, 2012. As of October 3, 2013:
http://www.swpc.noaa.gov/sww/SWW12_Poster_Abstracts.pdf

Davis, V. A., M. J. Mandell, and M. F. Thomsen, "Representation of the Measured Geosynchronous Plasma Environment in Spacecraft Charging Calculations, *J. Geophys. Res.*, Vol. 113, p. A10204. As of December 12, 2013:
http://onlinelibrary.wiley.com/doi/10.1029/2008JA013116/abstract

Dwork, C., "Differential Privacy," in M. Bugliesi, B. Preneel, V. Sassone, and I. Wegener, eds., *International Colloquium on Automata, Languages and Programming, ser. Lecture Notes in Computer Science*, Vol. 4052, Berlin, Heidelberg: Springer Berlin / Heidelberg, 2006, pp. 1–12. As of December 6, 2012:
http://dx.doi.org/10.1007/11787006_1

———, "Differential Privacy: A Survey of Results Theory and Applications of Models of Computation," in M. Agrawal, D. Du, Z. Duan, and A. Li, eds., *Theory and Applications of Models of Computation, ser. Lecture Notes in Computer Science*, Vol. 4978, Berlin, Heidelberg: Springer Berlin / Heidelberg, 2008, Ch. 1, pp. 1–19. As of December 6, 2012:
http://dx.doi.org/10.1007/978-3-540-79228-4_1

Dwork, C., F. McSherry, K. Nissim, and A. Smith, "Calibrating Noise to Sensitivity in Private Data Analysis Theory of Cryptography," in S. Halevi and T. Rabin, eds., *Theory of Cryptography, ser. Lecture Notes in Computer Science*, Vol. 3876, Berlin, Heidelberg: Springer Berlin / Heidelberg, 2006, Ch. 14, pp. 265–284. As of December 6, 2012:
http://dx.doi.org/10.1007/11681878_14

Eberhart, Jonathan, "ASAT Target Was Working Research Satellite," *Science News*, Vol. 128, September 28, 1985. As of December 12, 2013:
http://www.sciencenews.org/view/feature/id/188598

Fennell, J., H.C. Koons, M.S. Leung, and P.F. Mizera, *A Review of SCATHA Satellite Results: Charging and Discharging*, El Segundo: Aerospace Corporation, ADA158680, 1985. As of December 6, 2012:
http://www.dtic.mil/cgi-bin/GetTRDoc?AD=ADA158680

Gorney, D., and H. Koons, "Spacecraft Environmental Anomalies Expert System," paper presented at 28th Aerospace Sciences Meeting, AIAA, Reno, Nevada, January 8–11, 1990, Washington, D.C.: American Institute of Aeronautics and Astronautics, doi: 10.2514/6.1990-175, 1990. As of October 3, 2013:
http://arc.aiaa.org

Green, Janet, et al., physicist, National Oceanographic and Atmospheric Administration National Geophysical Data Center, discussion with the author and other physicists, Boulder, Colo., August 22, 2012a.

Green, Janet, William Murtagh, William Denig, Juan Rodriguez, Terry Onsager, J. Shoup, Jeff Stankiewicz, and Joe Kunches, "Report on the Satellite Anomaly Mitigation Stakeholders Meeting," paper presented at NOAA Space Weather Workshop, Boulder, Colo., National Oceanographic and Atmospheric Administration, April 23, 2012b. As of December 6, 2012: http://www.swpc.noaa.gov/sww/SWW_2012_Presentations/Tuesday_Morning_Satellites/SWWtalkfinaljgreen2.pptx

Grego, Laura, "A History of Anti-Satellite Programs," Cambridge, Mass.: Union of Concerned Scientists, January 2012. As of December 6, 2012: http://www.ucsusa.org/assets/documents/nwgs/a-history-of-ASAT-programs_lo-res.pdf

Goldreich, Oded, Sylvio Micali, and Avi Wigderson, "How to Play Any Mental Game," STOC '87, 1987, pp. 218–229. As of October 3, 2013: http://doi.acm.org/10.1145/28395.28420

Gussenhoven, M. S., D. A. Hardy, F. Rich, W. J. Burke, and H. C. Yeh, "High-Level Spacecraft Charging in the Low-Altitude Polar Auroral Environment," *J. Geophys. Res.*, Vol. 90, No. A11, 1985, pp. 11009–11023. As of October, 2013: http://dx.doi.org/10.1029/JA090iA11p11009

Heffetz, Ori, and Katrina Ligett, "Privacy and Data-Based Research," Social Science Research Network Working Paper Series, 2013. As of October 3, 2013: http://ssrn.com/abstract=2324830

Hoerlin, Herman, "United States High-Altitude Test Experiences: A Review Emphasizing the Impact on the Environment," LA-6405, Los Alamos, N.M.: Los Alamos Scientific Laboratory, October 1976. As of December 6, 2012: http://www.fas.org/sgp/othergov/doe/lanl/docs1/00322994.pdf

Hoffer, David, senior satellite analyst, Atrium Space Insurance Consortium, telephone communication with the authors, August 29, 2012a.

Hoffer, David, senior satellite analyst, Atrium Space Insurance Consortium, discussion with the authors, November 12, 2012b.

Iannotta, Becky, and Tariq Malik, "U.S. Satellite Destroyed in Space Collision," February 11, 2009. As of October 3, 2013: http://www.space.com/5542-satellite-destroyed-space-collision.html

Kivelson, Margaret G., and Christopher T. Russell, *Introduction to Space Physics*, Cambridge University Press, 1995.

Klanowski, Peter C., *Satellite News Digest: Satellite Outages and Failures*, web pages, 2012. As of November 17, 2012: http://www.sat-index.co.uk/failures/

Lam, H.-L., D. H. Boteler, B. Burlton, and J. Evans, "Anik-E1 and E2 Satellite Failures of January 1994 Revisited," *Space Weather*, Vol. 10, S10003, doi:10.1029/2012SW000811, 2012.

Lindell, Y., and B. Pinkas, "Secure Multiparty Computation for Privacy-Preserving Data Mining," *Journal of Privacy and Confidentiality*, Vol. 1, No. 1, pp. 59–98, 2009. As of December 6, 2012:
http://eprint.iacr.org/2008/197.pdf

Lohmeyer, W. Q., and K. Cahoy, "Space Weather Radiation Effects on Geostationary Satellite Solid-State Power Amplifiers," *Space Weather*, Vol. 11, No. 8, 2013, pp. 476–488. As of October 3, 2013:
http://dx.doi.org/10.1002/swe.20071

LWS Geospace Project Office, *An Evaluation of Continued SAMPEX Data Collection to the Goals of the Living With a Star Geospace Program*, NASA Living With a Star Geospace Project Office, 2003. As of December 6, 2012:
http://lwstrt.gsfc.nasa.gov/mowg0504_sampex.pdf

Maxwell Technologies, "SCS750 Super Computer for Space Datasheet," No. 1004741, Rev. 7 2012. As of December 6, 2012:
http://www.maxwell.com/products/microelectronics/docs/scs750_rev7.pdf

Mazur, J. E., and T. P. O'Brien, comment on Ho-Sung Choi et al., "Analysis of GEO Spacecraft Anomalies: Space Weather Relationships," *Space Weather*, Vol. 10, S03003, doi:10.1029/2011SW000738, 2012.

Mikaelian, Tsoline, "Spacecraft Charging and Hazards to Electronics in Space," York University, May 2001. As of October 2013:
http://arxiv.org/pdf/0906.3884.pdf

Narayan, Arjun, and Andreas Haeberlen, "DJoin: Differentially Private Join Queries over Distributed Databases," in *Proceedings of the 10th USENIX conference on Operating Systems Design and Implementation*, OSDI '12, Berkeley, Calif.: USENIX Association, 2012, pp. 149–162. As of October 3, 2013:
http://www.cis.upenn.edu/~ahae/papers/djoin-osdi2012.pdf

NASA—*See* National Aeronautics and Space Administration.

National Aeronautics and Space Administration, *Frequently Asked Questions: Orbital Debris*, 2011a. As of October 24, 2013:
http://www.nasa.gov/news/debris_faq.html

———, "USA Space Debris Environment, Operations, and Policy Updates," paper presented at 48th Session of the Scientific and Technical Subcommittee on the Peaceful Uses of Outer Space, United Nations, February 7–18, 2011b. As of October 3, 2103:
http://www.oosa.unvienna.org/pdf/pres/stsc2011/tech-31.pdf

———, *IMAGE RPI Anomaly*, web page, 2012a. As of December 6, 2012:
http://image.gsfc.nasa.gov/rpi/rpi_anomalies.html

———, *Living With a Star (LWS)*, web page, 2012b. As of November 1, 2013:
http://lws.gsfc.nasa.gov

———, *Storms From the Sun*, web page, March 8, 2012c. As of November 1, 2013: http://www.nasa.gov/mission_pages/sunearth/news/storms-on-sun.html

———, *Solar Cycle Prediction*, web page, updated January 2, 2014. As of January 23, 2014: http://solarscience.msfc.nasa.gov/predict.shtml

National Oceanic and Atmospheric Administration, National Geophysical Data Center, *Online Publications*, undated a. As of November 1, 2013: http://www.ngdc.noaa.gov/stp/solar/onlinepubl.html

———, Space Weather Prediction Center, *NOAA Space Weather Scales*, undated b. As of November 1, 2013: http://www.swpc.noaa.gov/NOAAscales/

———, Space Weather Prediction Center, *Satellites and Space Weather*, undated c. As of November 1, 2013: http://www.swpc.noaa.gov/info/Satellites.html

National Research Council, Committee for the Assessment of NASA's Orbital Debris Programs, *Limiting Future Collision Risk to Spacecraft: An Assessment of NASA's Meteoroid and Orbital Debris Programs*, Washington, D.C.: The National Academies Press, 2011. As of December 6, 2012: http://www.nap.edu/openbook.php?record_id=13244

National Science Foundation, *Geospace Environment Modeling (GEM)*, 2012. As of November 1, 2013: http://www.nsf.gov/funding/pgm_summ.jsp?pims_id=5506&org=AGS

NOAA—*See* National Oceanic and Atmospheric Administration.

NRC—*See* National Research Council.

Noushkam, Nikki, senior principal engineer, Orbital Sciences Corporation, telephone communication with the author, October 26, 2012.

O'Brien, Paul, Douglas G. Brinkman, Joseph E. Mazur, Joseph F. Fennell, and Timothy B. Guild, *A Human-in-the-Loop Decision Tool for Preliminary Assessment of the Relevance of the Space Environment to a Satellite Anomaly*, Aerospace Corporation, ATR-2011(8181)-2, May 10, 2012.

O'Brien, Paul, Timothy B. Guild, Joseph E. Mazur, and Richard K. Lee, *Spacecraft Environmental Anomalies Expert System (SEAES)—IR&D Developments FY06 to FY09*, El Segundo, Calif.: Aerospace Corporation, ATR-2009(8073)-3, November 15, 2009.

O'Brien, Paul, Joseph E. Mazur, and Timothy B. Guild, *Recommendations for Contents of Anomaly Database for Correlation with Space Weather Phenomena*, El Segundo, Calif.: Aerospace Corporation, TOR-2011(3903)-5, November 10, 2011.

O'Brien, T. P., "SEAES-GEO: A Spacecraft Environmental Anomalies Expert System for Geosynchronous Orbit," *Space Weather*, Vol. 7, No. 9, 2009, p. S09003. As of October 3, 2013:
http://dx.doi.org/10.1029/2009SW000473

Ostrovsky, R., and W. E. Skeith III, "A Survey of Single-Database Private Information Retrieval: Techniques and Applications," in T. Okamoto and X. Wang, eds., *Public Key Cryptography, ser. Lecture Notes in Computer Science*, Vol. 4450, New York: Springer, 2007, pp. 393–411. As of December 6, 2012:
http://dl.acm.org/citation.cfm?id=1760599

Park, J., Y. J. Moon, and N. Gopalswamy, "Dependence of Solar Proton Events on Their Associated Activities: Coronal Mass Ejection Parameters," *J. Geophys. Res.*, Vol. 117, No. A8, 2012, p. A08108. As of December 12, 2013:
http://dx.doi.org/10.1029/2011JA017477

Robinson, P. A., *Spacecraft Environmental Anomalies Handbook*, Hanscom Air Force Base, MA. Air Force Geophysics Laboratory. 1989.

Rodriguez, J., et al., "New Space Weather Particle and Magnetic Field Products at NGDC," poster presented at NOAA Space Weather Workshop, Boulder, Colo., National Oceanographic and Atmospheric Administration, April 23, 2012. As of October 3, 2013:
http://www.swpc.noaa.gov/sww/SWW12_Poster_Abstracts.pdf

Romero, M. and Levy, L., Internal Charging and Secondary Effects, in The Behaviour of Systems in the Space Environment (Proc. NATO Advanced Study Institute 245, Kluwer Academic Publishers), R. N. DeWitt et al. (eds.), pp. 565–580, 1993.

Satellite Interference Reduction Group, homepage, undated. As of October 3, 2103:
http://satirg.org

Shiga, David, "Mysterious Source Jams Satellite Communications," *New Scientist*, January 26, 2007. As of Oct 29, 2013:
http://www.newscientist.com/article/dn11033-mysterious-source-jams-satellite-communications.html

Siceloff, Steven, "Shuttle Computers Navigate Record of Reliability," NASA, 2010. As of 10/18/2012:
http://www.nasa.gov/mission_pages/shuttle/flyout/flyfeature_shuttlecomputers.html

SIRG—*See* Satellite Interference Reduction Group.

Space Telescope Science Institute, "Hubble Space Telescope Primer for Cycle 21," Shireen Gonzaga, Susan Rose, et al., eds., December, 2012. As of December 6, 2012:
http://www.stsci.edu/hst/proposing/documents/primer/Ch_2_Systemoverview3.html

Speich, Dave, and Barbara Poppe, "Space Environment Topics SE-16: Satellite Anomalies," NOAA Space Environment Center, 2000. As of December 6, 2012: http://www.swpc.noaa.gov/info/Satellite.pdf

Washington Post, "Falun Gong Jams Official Chinese TV," *Chicago Tribune*, July 9, 2002. As of November 12, 2012: http://articles.chicagotribune.com/2002-07-09/news/0207090078_1_falun-gong-li-hongzhi-hong-kong-based-human-rights-group

Tschan, Christopher, "Defensive Counterspace (DCS) Test Bed (DTB) for Rapid Spacecraft Attack/Anomaly Detection, Characterization, and Reporting," paper presented at Ground System Architectures Workshop, El Segundo, Calif., February 21, 2001. As of November 11, 2012: http://sunset.usc.edu/events/GSAW/gsaw2001/SESSION3/Tschan.pdf

Tschan, Christopher, Christopher Bowman, and Duane DeSieno, "Creating a Comprehensive Feature Space Library for Machine Learning," paper presented at Infotech@Aerospace 2012, Garden Grove, Calif., AIAA, 2012. As of October 3, 2013: http://arc.aiaa.org/doi/pdf/10.2514/6.2012-2444

Vandenberg Air Force Base, Joint Functional Component Command for Space factsheet, March 15, 2013. As of November 1, 2013: http://www.vandenberg.af.mil/library/factsheets/factsheet.asp?id=12579

Wade, David, David Hoffer, and Robin Gubby, "Financial Impact of Space Weather Anomalies—An Insurer's Perspective," paper presented at NOAA Satellite Anomaly Mitigation Stakeholders Meeting, Boulder, Colo., April 23, 2012. As of December 6, 2012: ftp://ftp.ngdc.noaa.gov/STP/publications/sam/SAM_Hoffer.pdf

Webb, R. C., L. Palkuti, L. Cohn, and G. Kweder, "The Commercial and Military Satellite Survivability Crisis," *Defense Electronics*, Vol. 24, 1995.

Wertz, James R., and Wiley J. Larson, eds., *Space Mission Analysis and Design*, 3rd ed., Hawthorne, Calif.: Microcosm Press, Inc., 1999.

Wren, Gordon L., and Andrew J. Sims, "Surface charging of spacecraft in geosynchronous orbit," *The Behavior of Systems in the Space Environment*: Proceedings of the NATO Advanced Study Institute, 1993.

Wrenn, G.L., D. J. Rodgers, K.A. Ryden, "A solar cycle of spacecraft anomalies due to internal charging," *Annales Geophysicae*, Vol. 20, 3/2002, 2002, pp. 953–956.

Yao, Andrew, "Protocols for Secure Computations (Extended Abstract)," *FOCS '82*, 1982, pp. 160–164. As of October 3, 2013: http://dx.doi.org/10.1109/SFCS.1982.88

Yao, Andrew, "How to Generate and Exchange Secrets," *FOCS '86*, 1986, pp. 162–167. As of October 3, 2013: http://portal.acm.org/citation.cfm?id=1382944